NANOPOLYMERS AND NANOCOMPOSITES

NANOPOLYMERS AND NANOCOMPOSITES

By

Dr. Helan Hawart

2016

SBS Publishers & Distributors Pvt. Ltd.
New Delhi

ISBN 13 : 9789380090658

First Published in 2016

Published by:

SBS PUBLISHERS & DISTRIBUTORS PVT. LTD.

2/9, Ground Floor, Ansari Road, Darya Ganj,

New Delhi - 110002,

INDIA

Tel: 0091.11.23289119 / 41563911

Email: mail@sbspublishers.com

www.sbspublishers.com

Preface

Nanopolymer is a polymer-based composite with a complex architecture, high density and durability. Nanopolymer reinforced some components of plasma conduits in emergency repairs or hazardous conditions. A Nanocomposite is a composite material, in which one of the components has at least one dimension that is around 10–9 m. A Nanocomposite is a multiphase solid material where one of the phases has one, two or three dimensions of less than 100 nm, or structure having nano-scale repeat distance between the different phases that make up the material. It is consist of one or more discontinuous phases of distributed in one continuous phase, continuous phase is called "matrix", whereas discontinuous phase is called "reinforcement" or "reinforcing material". Nanocomposites are found in nature, for example in the structure of the abalone shell and bone. The use of nanoparticle-rich materials long predates the understanding of the physical and chemical nature of these materials. Nanocomposites are upcoming materials, which shows the great changes in all the industrial fields and it is also going to be economical barrier for developing countries as a tool of Nanotechnology.

Editor

Contents

Chapter 1

COMPUTATIONAL STRATEGIES FOR POLYMER DIELECTRICS DESIGN

C.C. Wang[a, b], G. Pilania[c], S.A. Boggs[b], S. Kumar[d], C. Breneman[e], R. Ramprasad[a, b]

[a] Department of Materials Science and Engineering, University of Connecticut, 97 North Eagleville Road, Storrs, CT 06269, USA

[b] Institute of Materials Science, University of Connecticut, 97 North Eagleville Road, Storrs, CT 06269, USA

[c] Materials Science and Technology Division, Los Alamos National Laboratory, Los Alamos, NM 87545, USA

[d] Department of Chemical Engineering, Columbia University, 500W. 120th St., New York, NY 10027, USA

[e] Rensselaer Exploratory Center for Cheminformatics Research and Department of Chemistry and Chemical Biology, Rensselaer Polytechnic Institute, Troy, NY 12180, USA

ABSTRACT

The present contribution provides a perspective on the degree to which modern computational methods can be harnessed to guide the design of polymeric dielectrics. A variety of methods, including quantum mechanical ab *initio* methods, classical force-field based molecular dynamics simulations, and data-driven paradigms, such as quantitative structure–property relationship and machine learning schemes, are discussed. Strategies to explore, search and screen chemical and configurationally spaces extensively are also proposed. Some examples of computation-guided synthesis and understanding of real polymer dielectrics are also provided, highlighting the anticipated increasing role of such computational methods in the future design of polymer dielectrics.

GRAPHICAL ABSTRACT

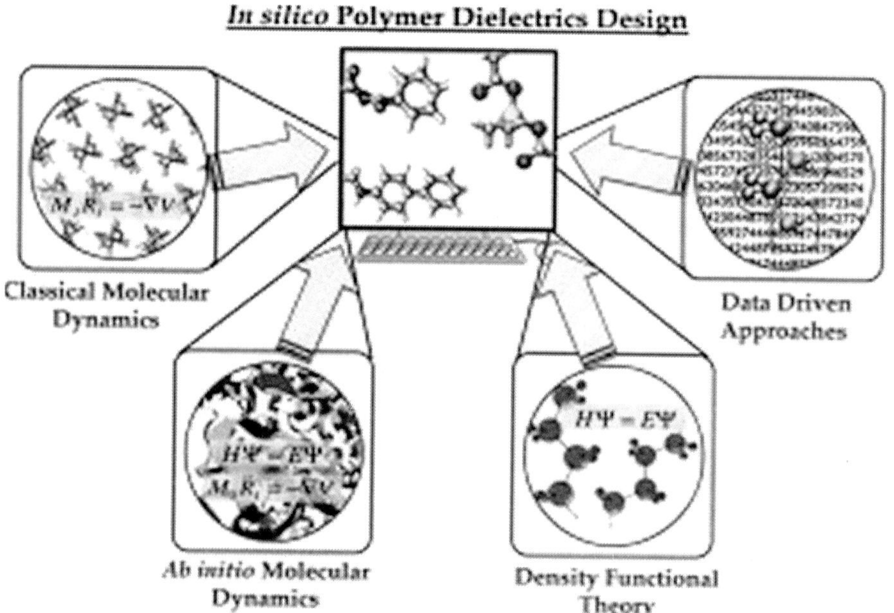

INTRODUCTION

Polymers offer a nearly infinite variety of material systems with diverse properties. Until recently, the formulation of polymers for specific applications was based on trial and error, guided by intuition. The purpose of the present contribution is to demonstrate the degree to which computational methods can guide the design of polymers, in the present case for dielectric applications [1], [2], [3], [4], [5], [6], [7], [8], [9], [10], [11] and [12], which require high dielectric constant, large band gap, high dielectric strength, low dielectric loss, and appropriate glass transition temperature and morphology.

At the most fundamental level, computational quantum mechanics, e.g., density functional theory (DFT), can be used to determine properties of dielectrics at the scale of a crystalline unit cell [13], [14], [15], [16] and [17]. Such properties include structural and thermodynamic details, reasonable estimates of the band gap, electronic dielectric constant, ionic dielectric constant, and intrinsic breakdown field [18], [19], [20], [21], [22], [23] and [24]. In addition, impurity states in the band gap caused by common chemical impurities can be computed [25], [26], [27] and [28]. Realistic models can also be developed for metal–polymer interfaces in order to predict charge injection characteristics.

Larger scale morphological features of polymers can be accessed practically at the present time only using molecular dynamics (MD) based on empirical interatomic potentials or force fields [29], [30], [31], [32] and [33]. Such simulations can predict crystal structure, semicrystalline morphology and provide rough estimates of glass transition temperature and dielectric loss, although the latter is presently limited to loss in the GHz range [34], [35], [36] and [37].

The above methods can be classified as "physics-based", as they are based on quantum mechanics, classical mechanics, and classical electromagnetism. An emerging class of methods, often referred to as "data-driven", use various forms of multivariate analysis on experimental or computational data, based on complex variables with a physical relationship to the properties being predicted [38], [39], [40], [41], [42], [43], [44], [45] and [46]. Such systems are "trained" on available data and then used to predict properties of interest for polymers for which data are not available. An example of such data-driven approaches is quantitative structure property relationships (QSPR) [47], [48], [49], [50] and [51], which can predict properties, such as glass transition temperature, melting temperature, etc., for which no fundamental approach is presently available.

In this contribution, we provide a perspective on the application of modern computational approaches to the design of polymeric dielectrics. Section 2 addresses functionalization of a well-understood polymeric dielectrics, such as polyethylene (PE) and polypropylene (PP), to enhance its dielectric response. Section 3 discusses approaches to the discovery of entirely new classes of polymer dielectrics, both organic and organometallic. Strategies discussed include exploration of large chemical spaces and efficient computation of some relevant properties. The proper starting point for such chemical space exploration is a quantum mechanics based method such as DFT, but the promising systems identified using quantum mechanics must be investigated using other methods that treat the morphology (e.g., force field based MD), glass transition temperature (QSPR), melting temperature (QSPR), etc. The screening of systems considered in Section 3 is based on dielectric constant and band gap. Appropriate values for these properties are "necessary but not sufficient" for a given application and, therefore, appropriate for first-level screening. After the first-level screening for these properties, promising candidates can be subjected to more detailed analysis for a broader range of properties required for a given application. These include factors that control the injection of charge carriers from a metal electrode into the polymer, formation of defect states in the polymer, factors that control the transport of charge carriers (e.g., scattering due to dipoles, impurities and phonons), dielectric breakdown, glass transition temperature, etc. Such aspects are discussed in Section 4. The role data-driven approaches, such as QSPR and, in particular, the emerging area of "machine learning", is surveyed in Section 5. We conclude with our outlook in Section 6.

MODIFICATION OF EXISTING POLYMER DIELECTRICS

A good place to start in applying computational methods to design new materials is to understand or identify the factors that lead to surprising properties when a well-studied material is modified creatively. Polypropylene (PP), in its biaxially oriented form, is the most common polymer dielectric for high energy density capacitors as a result of its high dielectric breakdown strength, low dielectric loss, and good clearing characteristics; however, its dielectric constant is only 2.2 [52], [53] and [54]. Increasing the dielectric constant while maintaining or improving other properties, such as operating temperature range, is highly desirable. Chung has developed an –OH functionalized PP which doubles the dielectric constant and maintains relatively low loss with only 4.2 mol% –OH (Fig. 1) [55], [56] and [57]. The experimental data also imply that the PP-OH contains roughly 0.5 water molecule per –OH moiety on the chain and that the –OH groups are hydrogen-bonded in pairs.

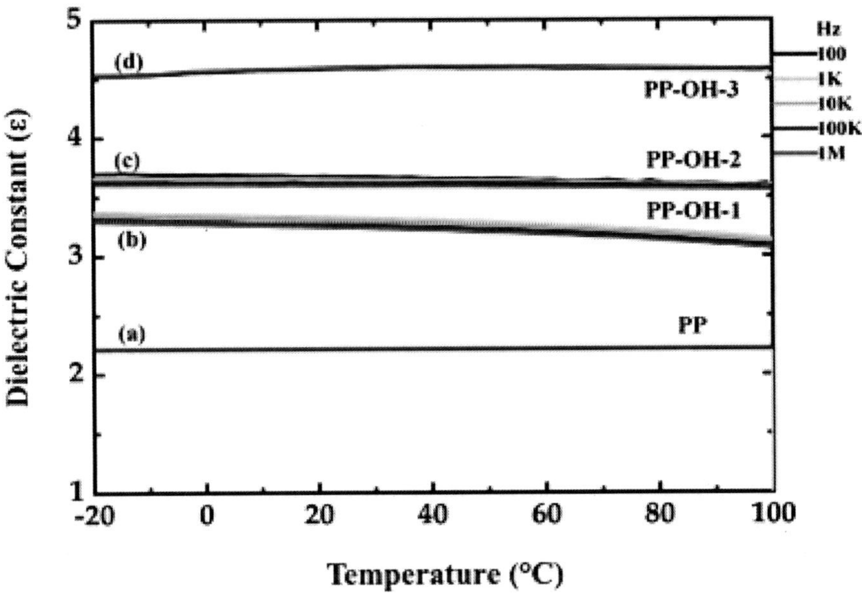

Figure 1. Dielectric constants (vs. frequency and temperature) for (a) polypropylene (PP) and three hydroxyl (–OH) functionalized PP (PP–OH) copolymers containing (b) 0.7, (c) 1.8, and (d) 4.2 mol% OH content. Replotted from data in Ref. [55].

DFT computations can investigate the role played by –OH functional groups and the trapped moisture [59]. Given that DFT computations are time-intensive, two short-chain polyethylene oligomers (o-PE), each with 11 carbon atoms, were considered. This o-PE system is chemically similar to PP, but much simpler and smaller in size. The two chains were arranged in a head-to-tail configuration, and an H chain end atom was substituted by an –OH group (o-PE–OH). Additionally, to explore the effect of water molecules, o-PE–OH with one and two water molecules (o-PE–OH–H_2O and o-PE–OH–$2H_2O$, respectively) were studied. Fig. 2 shows snapshots of the optimized geometries. The average value of the total dielectric constant, i.e., the trace of the dielectric constant tensor, including both ionic and electronic contributions, for the o-PE, o-PE–OH, o-PE–OH–H_2O, and o-PE–OH–$2H_2O$ systems were, respectively, 2.4, 3.3, 3.7, and 4.3 (Fig. 3) [59]. The determination of the dielectric constant of such chain systems was performed using a recently developed method aimed at precisely these situations

[19]. For pure PE, the dielectric constant, which is almost entirely due to electronic contributions, is close to the accepted experimental value. The increase in the dielectric constant due to the -OH groups (without and with water molecules) is caused entirely by the ionic contributions. The proper account of van der Waals (vdW) interactions [60] and [61], secondary bonding phenomena such as H-bonding, etc., is captured adequately in the DFT computations. For instance, Fig. 2(d) shows the relaxed geometry of the o-PE-OH-$2H_2O$ system, where the formation of an H-bonded ring containing two -OH groups and $2H_2O$ molecules, is evident.

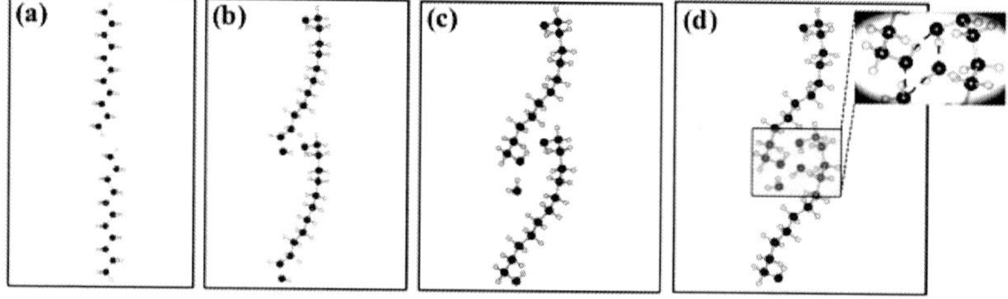

Figure 2. Optimized structures of (a) o-PE, (b) o-PE-OH, (c) o-PE-OH-H_2O, and (d) o-PE-OH-$2H_2O$. Black, white, and red spheres represent C, H, and O atoms, respectively. The inset shows a typical hydrogen bonded ring.

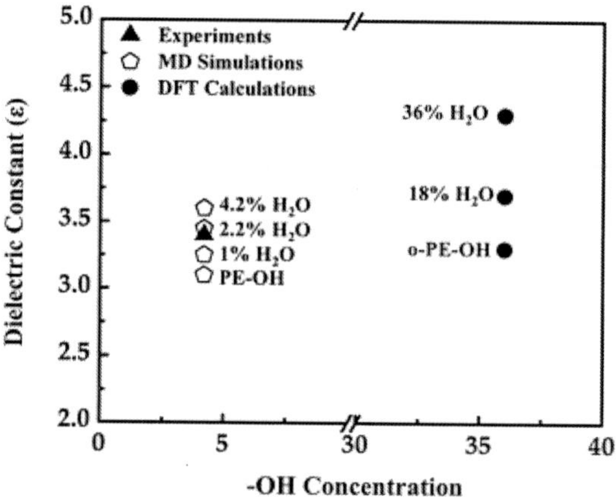

Figure 3. Comparison of the dielectric constant of hydroxyl functionalized polyethylene (PE-OH), obtained from experiment, MD simulations and DFT calculations.

While the DFT computations provide some understanding of the role played by -OH functional groups and the accompanying water molecules, these data are qualitative. The DFT analysis ignores complexities associated with morphological variations of PE when functionalized with -OH, and owing to computational considerations, the concentration of -OH modeled is much greater than in experiments. Large-scale MD simulations based on force fields are required to address morphological aspects of the system at realistic -OH concentrations. Recent force-field based MD simulations have extended the DFT work discussed above [62]. In the MD simulations, PE–OH with 4.2 mol% -OH groups was considered, and varying amounts of water were added into PE-OH system. The morphology of -OH groups, trapped water and H-bonding was well captured, insofar as can be concluded by comparisons with measured infrared spectra. In addition, the MD simulations indicate that the -OH groups tend to collect at amorphous–crystalline interfaces, likely as a result of -OH groups "expelled" from the crystalline regions during formation. These simulations also imply that hydrogen bonding of the -OH groups into pairs is essential to maintaining low dielectric loss. Fig. 3 summarizes data for the dielectric constant derived from the MD simulations, in addition to the results from DFT computations and experiment. The addition of water results in a significant increase in dielectric constant. For ~0.5 water per -OH group, the MD data for dielectric constant match closely the experimental value of ~3.4, and larger amounts of water, so long as they are "trapped" by the -OH groups much like in Fig. 2(d), increase the dielectric constant further without increasing loss.

The DFT, MD and experimental data, collectively, have allowed us to understand the factors that lead to a significant increase in the dielectric constant of a saturated hydrocarbon due to the incorporation of a small amount of -OH groups. These findings are significant, as they imply a path toward tunable control of dielectric properties of polyolefins through functionalization. While the analysis presented underlines the complexities that can result from incorporation of a functional group, it also demonstrates the utility of computational strategies for systematic studies of other functional groups.

PATHWAYS TO THE DISCOVERY OF NEW POLYMER DIELECTRICS

While modifying an existing polymer to enhance its properties is a promising approach and offers a "risk mitigation" strategy, identifying or discovering entirely new classes of polymers can be transformative but requires systematic schemes to explore the polymer chemical space. Fig. 4 portrays one such scheme in which single polymeric chains consist of four distinct building blocks drawn from a pool of possibilities in a combinatorial manner [19] and [20]. Depending on the pool of blocks, various polymer classes can be studied. The combinatorial explosion produced by this strategy is both a blessing and a curse. For example, 10 possibilities for each of the 4 blocks in the repeat unit result in a total of 10,000 cases, although translational and inversion invariances, and other considerations such as chemical intuition, can reduce this number considerably.

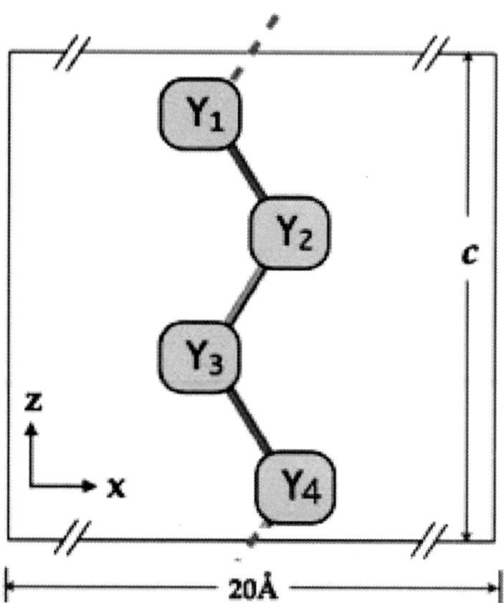

Figure 4. Schematic representation of the polymeric chain model. c is the lattice constant along the polymer chain direction that was allowed to relaxed along with the internal coordinates. Four blocks Y_1, Y_2, Y_3, and Y_4 can be filled up by different motifs.

New Organic Polymers

We first consider a set of organic building blocks, for example, $-CH_2^-$, $-NH^-$, -C(double bond; length as m-dashO)$^-$, $-C_6H_4^-$ (benzene), $-C_4H_2S^-$ (thiophene), -C(double bond; length as m-dashS)$^-$, and $-O^-$ [63] and [64], which are common in polymer backbones and various combinations of which form traditional polymers, including polyesters, polyamides, polyethers, polyureas, etc. Accounting for translational and inversion symmetries, and removing unstable systems, leaves 267 symmetry-unique systems.

The two most important properties of dielectrics for most applications are dielectric constant and band gap. Hence, we focus on these two properties in this section. With the aid of a recently developed method which determines the dielectric constant of such single chain systems [19], high throughput DFT computations can be used to obtain the band gap and dielectric constant of the 267 polymer systems. Fig. 5(a)–(c), respectively, shows the relationship between the electronic, ionic and total dielectric constant and band gap for the 267 polymer systems.

A near perfect inverse Pareto optimal front relationship between the band gap and the electronic dielectric constant can be seen from Fig. 5(a), which imposes a theoretical limit on the electronic part of the dielectric constant as a function of band gap, a limit that can be understood by regarding the electronic part of the dielectric response as a sum over electronic transitions from occupied to unoccupied states. On the other hand, the ionic dielectric constant is not correlated with the band gap, as seen from Fig. 5(b). The ionic contribution is determined by the infrared active zone center phonon modes (i.e., the modes that display a time-varying dipole moment) [65] and [66].

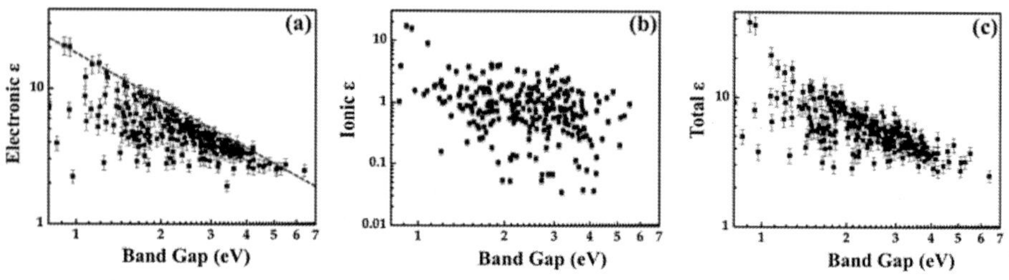

Figure 5. (a) Electronic, (b) ionic and (c) total dielectric constant (ε) as a function of the band gap, for the class of organic polymers considered, computed using DFT within the single-chain approach. The building blocks were drawn from the following pool of possibilities: $-CH_2^-$, $-NH^-$, -C(double bond; length as m-dashO)$^=$, $-C_6H_4^-$ (benzene), $-C_4H_2S^-$ (thiophene), -C(double bond; length as m-dashS)$^-$, and $-O^-$. The axes are in logarithmic scale.

The ionic dielectric constant can thus be exploited to increase the total dielectric constant without compromising the band gap. Fig. 5(c) shows the variation of the total dielectric constant with the band gap. Plots such as Fig. 5(c) provide us with a "map" of the achievable combination of properties within the chemical space explored. Capacitive energy storage and some electronics applications, e.g., gate insulations, could draw from the large dielectric constant and moderate band gap region of this plot. Desirable material properties are likely to be associated with certain building block structures. Such observations could accelerate the design process by identifying correlations between material properties and specific building blocks structures. Indeed, guided by the DFT results, a few polyureas and polyurethanes containing $-NH^-$, -C(double bond; length as m-dashO)–, $-C_6H_4^-$, and $-O^-$ building blocks have been synthesized using step polymerization [64].

Once a set of promising polymers has been identified, their crystal structures and morphologies must be investigated. If inter-atomic potentials are available to handle the systems, MD simulations can be used to determine their crystal structure and morphology, for example, using a melt and quench approach [34], [35], [36] and [37]. An additional option, especially if force fields are not available for the new identified systems, is to use 3-dimensional structure searching schemes [67], [68], [69], [70], [71] and [72] to determine the ground state structures based on DFT.

Polymer blocks containing group 14 elements

The strategy described above can be extended by considering building units based on non-carbon elements from Group 14 of the periodic table, i.e., Si, Ge and Sn. Replacing C with Si, Ge or Sn offers the opportunity to manipulate the band gap and the electronic part of the dielectric constant through control of σ conjugation along the chain, and to manipulate the ionic part of the dielectric constant through control of dipole moments. Substituting C with Si, Ge and/or Sn ensures chemical compatibility by preserving the local chemical environment and bonding. The dipole moments of each building block can be enhanced by introducing small atoms with high electronegativity such as F and Cl to the side chain. We thus consider polymers with building block units represented as $-XY_2^-$ in this section, where X = C, Si, Ge, Sn and Y = H, F, Cl.

Before exploring the single chain systems based on the above combinatorial exercise, we investigate $-XY_2^-$ "homopolymer" crystals to see if any pattern emerges. Group 14 systems display rich chemistry and crystallize in forms with differing coordination geometries. While C always prefers a 4-fold coordination environment, Ge favors a 5-fold environment, and Sn can occur in 6 or 7 fold coordination geometries [19] and [73]. Fig. 6 presents dielectric constant and band gap data for homopolymer ground state geometries. The electronic part of the dielectric constant tends to vary inversely with the band gap, as expected, while the ionic contribution to the dielectric constant is negligible for the C based systems, low for the Si based systems, but quite high for the Ge and Sn containing polymers. Fig. 6 also reveals that $-XY_2^-$ homopolymers with Ge and Sn backbone have large dielectric constants but smaller band gaps than C and Si-based polymers [19] and [73]. Pure PE has a large band gap but relatively small dielectric constant. A natural next step is to "mix" PE with the $-XY_2^-$ homopolymers to identify compositions which span a large range of band gap and dielectric constant. Such a combinatorial exploration based on the scheme presented in Fig. 4 involving 7 building block possibilities (namely, $-CH_2^-$, $-SiF_2^-$, $-SiCl_2^-$, $-GeF_2^-$, $-GeCl_2^-$, $-SnF_2^-$ and $-SnCl_2^-$) was recently undertaken [20] and [38]. Results for dielectric constant vs. band gap for the 175 such single chain polymers are shown in Fig. 7. While PE has the greatest calculated band gap of the systems explored, addition of Group 14 elements leads to progressive decrease in the band gap and increase in the electronic dielectric constant. The total dielectric constant of the Group 14 element-based polymers spans over a large range between 2.5 and 47, with the smallest and largest values corresponding to $-(CH_2)_4^-$ and $-CH_2^-(SnF_2)_3^-$, respectively. As the backbone atoms vary from C to Sn with all other units in the chain are held fixed, both the electronic and total dielectric constant increase, as the band gap decreases, probably because $-SnF_2^-$ has the largest dipole moment and the Sn–Sn bond rotation has the lowest barrier among all the X–X (with X or X = C, Si, Ge, and Sn). Compared with organic polymers, the Group 14 element-based hybrid polymers, especially those involving Sn, can achieve greater dielectric constants without large reductions of the band gap, which makes them attractive candidates for high dielectric constant polymeric dielectrics [20] and [38].

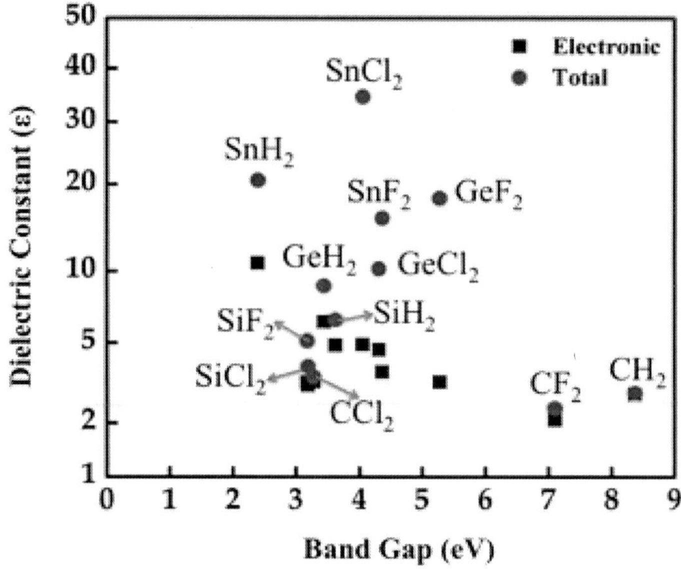

Figure 6. Dielectric constant and band gap results for $-XY_2^-$ homopolymer crystals at their respective ground state structure (where X = C, Si, Ge, Sn and Y = H, F, Cl).

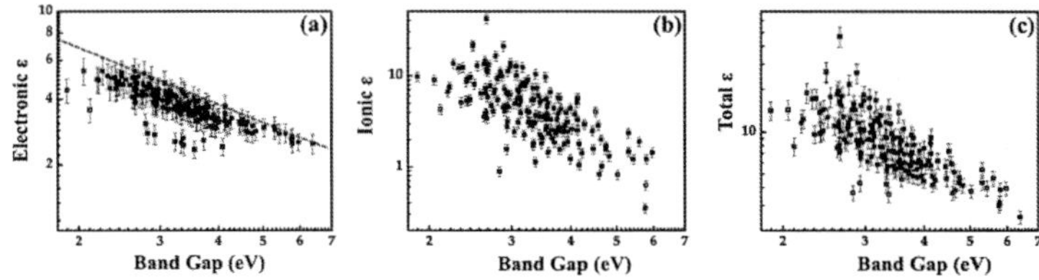

Figure 7. (a) Electronic, (b) ionic and (c) total dielectric constants (ε) as a function of the band gap, for the class of polymers containing Group 14 elements, computed using DFT within the single-chain approach. The building blocks were drawn from the following pool of possibilities: $-CH_2^-$, $-SiF_2^-$, $-SiCl_2^-$, $-GeF_2^-$, $-GeCl_2^-$, $-SnF_2^-$ and $-SnCl_2^-$. The axes are in logarithmic scale.

ADVANCED SCREENING STRATEGIES

The above discussion dealt largely with dielectric constant and band gap. While these two properties are important, many other factors are relevant to practical applications. Thus down-selection of promising candidates requires estimating a range of other parameters. In the following, we describe briefly a selection of such factors, to highlight the present state-of-the-art in the computation of material properties, and the challenges that need to be overcome.

Electron/Hole Injection Barriers At Metal–Polymer Interfaces

Electronic and capacitor applications require the polymer dielectric to contact a metal electrode. Such an interface sometimes dominates the performance of polymer dielectrics, as charge injection from the metal electrode to the polymer is the primary source of charge during conduction. The property that determines charge injection at interfaces is the Schottky barrier heights for electron and hole injection, which can be computed using modern DFT. Such issues have been encountered in past studies of metal–dielectric or metal–semiconductor interfaces within the context of (inorganic) semiconductor devices [74], [75], [76] and [77]. While methods developed earlier can be used within the context of metal–polymer interfaces, the primary impediment to doing so is the complex nature of the metal–polymer interface at the atomic-level, which is far from ideal or "abrupt", unlike interfaces typically encountered by the semiconductor community. Future efforts to compute barrier heights must break new ground in terms of realistic models of interfaces, perhaps through a combination of DFT and force-field based computations.

Defects in Insulating Polymers

Chemical impurities in polymer dielectrics introduce localized states in the band gap which can act as traps and thereby affect conduction and high field aging (i.e., the gradual degradation of the dielectric). Previous studies based on DFT provide an understanding of impurity states in the band gap resulting from chemical impurities and conformational distortion [25], [26], [27] and [28]. While states caused by the latter lie very near to the band edges (the conduction band in particular), those caused by the former tend to be separated from the valence and conduction band edges by $\gtrsim 1$ eV, as shown in Fig. 8. States close to the valence band edge are normally occupied and represent hole traps, while those close to the conduction band are normally empty and represent electron traps. The hole (electron) traps fill when positive (negative) space charge is injected into the material at high field.

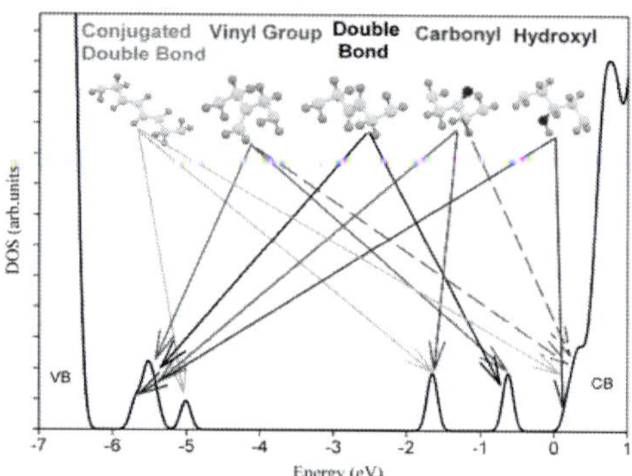

Figure 8. Density of states for polyethylene showing the impurity states created by a range of common chemical impurities. The zero of energy is referenced to the vacuum level, which is about 0.3 eV below the conduction band minimum of crystalline polyethylene [25].

Polymer aging occurs over time as a result of thermal, oxidative, mechanical, electrical, etc., effects, and can be caused by the breaking of primary chemical bonds in the chain backbone or side groups. Aging can also cause conformational changes which result in reduced crystallinity. Experimental evidence suggests that high field aging occurs as a result of recombination of electrons and holes as the opposite space charge fields pass through each other [78]. Under repetitive DC charging at sufficiently high fields, carrier recombination can occur at electrodes as a result of voltage cycling, and carrier recombination in the bulk occurs during steady state voltage. In either case, carrier recombination releases sufficient energy to break chemical bonds, which, in the presence of oxygen, can result in carbonyl formation, increased impurity states in the band gap, and enhanced conduction to form a positive feedback loop which eventually leads to breakdown.

A microscopic mechanism governing the initiating step in the high-field aging of crystalline polyethylene has been recently proposed based on DFT calculations and ab initio MD simulations [79]. The study assumed that electrons, holes, and excitons are present in the system. The key finding of this work is that the presence of triplet excitons can be damaging. The electron and hole states of the exciton localize on a distorted region of polyethylene and weaken nearby C–H bonds facilitating C–H bond scission. The estimated barrier to cleavage of the weakened C–H bonds is comparable to the thermal energy, suggesting that this mechanism may be responsible for the degradation of polyethylene when placed under electrical field, e.g., in high-voltage cables. Fig. 9 shows a snapshot of this process during the course of a DFT-based MD simulation. Laurent et al. have quantified these phenomena experimentally, to some degree, based on electroluminescence measurements [80].

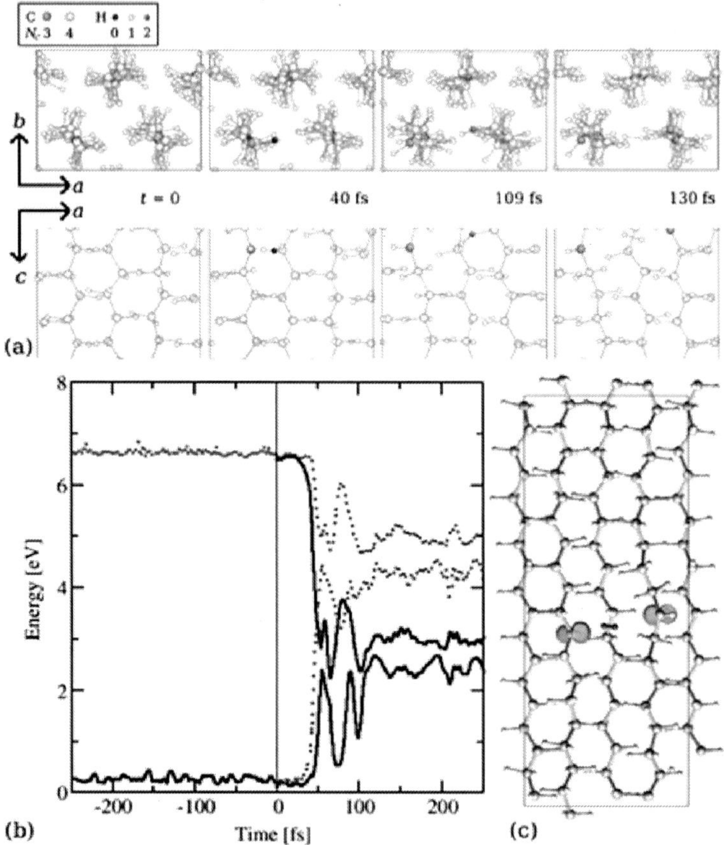

Figure 9. (a) Snapshots taken at time t during ab initio MD simulations of polyethylene, where t = 0 refers to the point at which the exciton was injected. Atoms are shaded according to coordination number NC. The snapshots illustrate C–H bond scission and the subsequent H abstraction. (b) Evolution of the highest occupied and lowest unoccupied energy eigenvalues of the system for 250 fs prior to (t < 0) and after (t > 0) exciton injection. The zero of energy is the value of the highest occupied PE eigenvalue at T = 0. Prior to exciton injection, each of the two levels shown is doubly occupied (the colors shown correspond to spin up states). Upon triplet exciton injection, both hitherto spin-degenerate states split into an occupied spin up (full black lines) and an unoccupied spin down (dotted blue lines) state. (c) Magnetization density plot of the system at t = 121 fs. The magnetization is tightly localized on the –CH^- groups along the main chains, indicating that these groups have unpaired electrons. The H_2 dimer is shown in deep red for clarity. Reprinted with permission from Ref. [79].

Dielectric Breakdown

The dielectric breakdown field of a capacitor dielectric is correlated strongly with its energy density (i.e., its energy storage capacity). Although extrinsic factors, e.g., chemical impurities, nano-sized cavities, etc., are important in determining breakdown field for polymer-based dielectrics, a firm theoretical understanding of the factors governing the intrinsic breakdown field is an essential first step in developing a model relevant to practical materials. Theories of intrinsic breakdown were developed in the 1930s by Von Hippel [81], Zener [82], and Frohlich [83]. Recently an approach to treat intrinsic breakdown using DFT in the average electron model of Von Hippel has been developed [21] and [22]. The principal phenomenon governing the intrinsic breakdown field is scattering of conduction-band electrons by phonons, i.e., crystal lattice vibrations. A simple model which reflects the balance between the average energy gain of electrons from the electric field and average energy loss through phonon scattering provides a basis for computing intrinsic breakdown. Upon application of a field smaller than the breakdown field, the average electron energy increases to some stable value at which the rate of energy gain from the field is balanced by energy loss to the lattice. Von Hippel "low energy criterion" defines the breakdown field as that at which the average rate of energy gain is greater than the average rate of energy loss for all energies up to the threshold for impact ionization. At this point, a rapid increase in the density of conduction electrons occurs, which we assume is sufficient to cause breakdown.

Fig. 10 compares the calculated intrinsic breakdown with the maximum experimentally observed breakdown field for a range of ionic and covalent materials [21] and [22]. Good agreement is achieved in that the computed breakdown field tracks the upper bound of the measured intrinsic breakdown field. Understanding of intrinsic breakdown is reasonably good for crystalline materials, but computation of engineering breakdown fields requires accounting for effects of morphology, chemical impurities, and defects such as nanocavities.

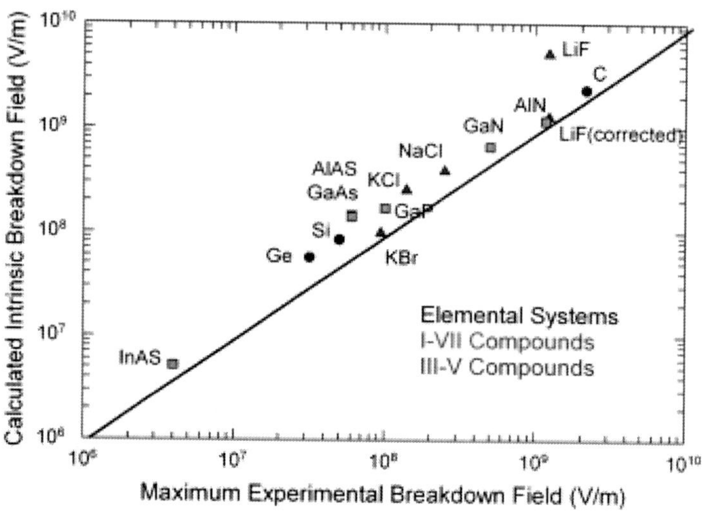

Figure 9. (a) Snapshots taken at time t during ab initio MD simulations of polyethylene, where t = 0 refers to the point at which the exciton was injected. Atoms are shaded according to coordination number NC. The snapshots illustrate C–H bond scission and the subsequent H abstraction. (b) Evolution of the highest occupied and lowest unoccupied energy eigenvalues of the system for 250 fs prior to (t < 0) and after (t > 0) exciton injection. The zero of energy is the value of the highest occupied PE eigenvalue at T = 0. Prior to exciton injection, each of the two levels shown is doubly occupied (the colors shown correspond to spin up states). Upon triplet exciton injection, both hitherto spin-degenerate states split into an occupied spin up (full black lines) and an unoccupied spin down (dotted blue lines) state. (c) Magnetization density plot of the system at t = 121 fs. The magnetization is tightly localized on the –CH– groups along the main chains, indicating that these groups have unpaired electrons. The H_2 dimer is shown in deep red for clarity. Reprinted with permission from Ref. [79].

Glass transition temperature

We now move from electrical properties to an important physical property, the glass transition temperature, Tg. Many properties change dramatically when an amorphous polymer undergoes glass transition. For instance, the dielectric loss of a polar material normally increases dramatically as the polymer goes through its Tg. At temperatures below Tg, polymers are hard and glassy because motion of polymer chains is restricted to local vibration. When the temperature goes to above Tg, chains have greater mobility and the polymer softens. Various factors affect the Tg of polymers, e.g., molecular weight and the chemical structure.

While Tg can be estimated using MD simulations, the cooling and heating rates that can be simulated are much greater than reality. Thus, Tg estimated using MD tends to carry large error bars. A second alternative to computing Tg is provided by Quantitative Structure–Property Relationship (QSPR) methods [47], [48], [49], [50] and [51]. The philosophy underlying a QSPR study is to relate structural information through descriptors to macroscopic properties. The ability of QSPR to predict the Tg of homopolymers (as well as the polarizabilities, dielectric constants, and HOMO–LUMO gaps of some polymers and functional groups) has already been demonstrated [39]. The polymers were represented using the repeat unit structure, end-capped with a monomer. After selecting the appropriate descriptors, statistical methods were applied to generate linear or nonlinear models. Fig. 11 compares experimental and predicted Tg for two QSPR models and demonstrates the predictive ability of the QSPR models for Tg. The wide structural diversity of the polymers proves that this is a general method for predicting Tg of non-cross-linked polymers which can be used reliably during screening studies.

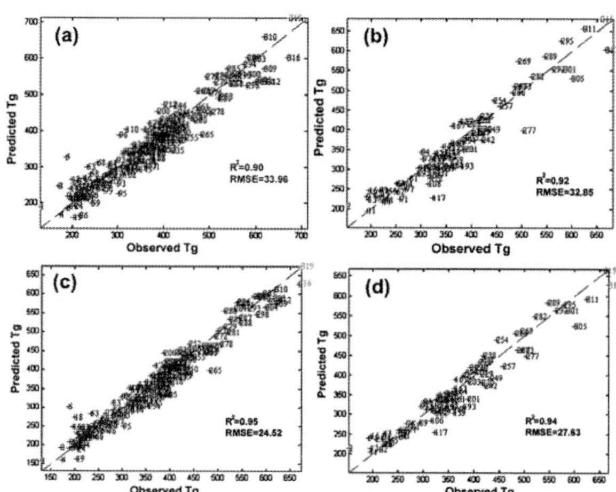

Figure 11. Partial least square (PLS) model for Tg ((a) training set and (b) test set) and Kernel PLS model for Tg ((c) training set and (d) test set). Reprinted with permission from Ref. [39].

While QSPR methods cannot generally duplicate the accuracy of physics-based calculations, they are capable of quickly providing remarkably useful results within well-defined domains of applicability when the models are trained using appropriate physics-based descriptors, and are subject to "best practices" model building and validation criteria. Often, the most important determinant for the usefulness of a QSPR model in any design project is whether the domain of applicability of the model can be defined.

ADVANCED SEARCHING STRATEGIES

DFT-based strategies are limited to polymers based on short, periodic chains. For longer chains required to screen larger regions of compositional space, the computational cost associated with DFT rises rapidly. Furthermore, as the system size increases, the number of candidates within the system grows exponentially, which leads to combinatorial explosion. Take the Group 14 element-based hybrid polymers for example. Doubling the supercell size along the chain direction to include 8 distinct building units in a periodic repeating cell results in a total of 29,365 symmetry unique systems. Clearly, exploration of such a vast chemical space using present first principles based approaches is impractical. A new approach is needed for this large class of systems.

To effect such large scale explorations, a machine learning approach has been developed and applied to the Group 14 element-based hybrid polymers [38]. The initial dataset was generated using the high throughput DFT calculations. To validate the machine learning algorithm, five properties in addition to dielectric constant and band gap were computed, viz., atomization energy, formation energy, c lattice parameter, spring constant, and electron affinity. Fig. 12 shows the agreement between predictions of the model and DFT computations which were separated randomly into "training" and "test" sets, the former used in model development and the latter for validation. Several chains composed of 8-block repeat units, in addition to the 4-block systems portrayed in Fig. 4, were considered. As can be seen, the agreement between the DFT and the learning schemes is uniformly good for all eight properties across the 4-block training and test set, as well as the somewhat out-of-sample 8-block test set. The fidelity of the model predictions is impressive, given that these calculations take a very small fraction of the time required for a typical DFT computation.

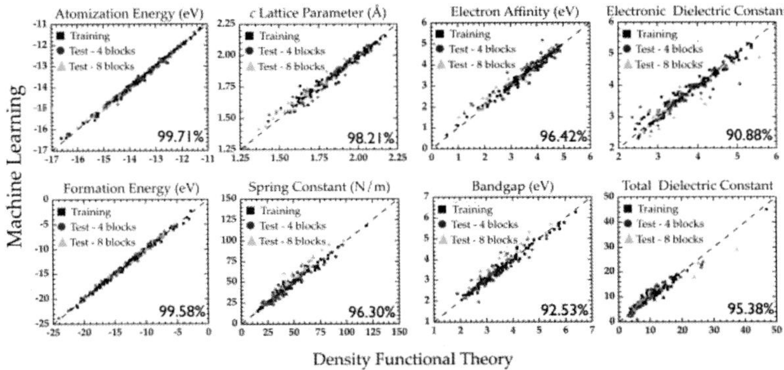

Figure 12. Parity plots comparing property values computed using DFT against predictions made using learning algorithms trained using chemo-structural fingerprint vectors. Pearson's correlation index is indicated in each of the panels to quantify the agreement between the two schemes [38].

While the good agreement between the machine learning and DFT data is gratifying, the real power of this approach lies in the possibility of exploring a much larger chemical space than is practical using DFT calculations or experiments. As mentioned above, expanding the 4-block Group 14 element-based hybrid polymers to 8-blocks results in 29,365 symmetry unique systems. The machine learning approach makes practical a study of this magnitude. Fig. 13 shows the predictions of band gap, electronic and total dielectric constant for the 29,365 systems. The inverse correlation of the band gap with the electronic dielectric constant is confirmed once again from Fig. 13(a). The rough inverse correlation between total dielectric constant and band gap seen in Fig. 13(b) is surprising, although clear deviations from this inverse behavior are seen. This extensive search facilitates identification of candidate polymer dielectrics for various applications. For capacitor applications, a search for polymers with high dielectric constant and large band gap leads to systems in the top part of Fig. 13(b), corresponding to the 'deviations' from the inverse correlation and indicated by a circle in Fig. 13(b). These are systems that contain two or more contiguous SnF2 units, but with an overall CH2 mole fraction of at least 25%. Such organo-tin systems may be appropriate for polymer dielectrics. By learning effectively from the available DFT data, the machine learning approach facilitates searching a much larger scale polymer chemical space efficiently. With the aid of this approach, the discovery of new polymer dielectrics can be accelerated significantly.

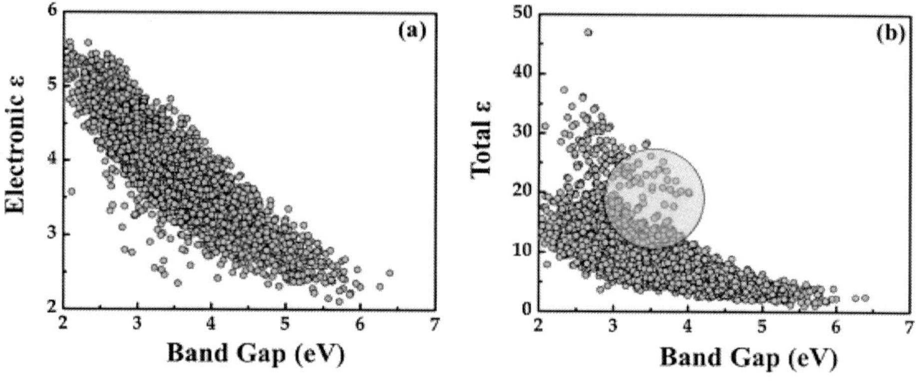

Figure 13. Machine learning predicted (a) electronic and (b) total dielectric constant as a function of the band gap for the 8-block systems. The DFT results for the 4-block systems (gray circles) are also shown in the figure as comparison. Replotted from data in Ref. [38].

CONCLUSION

In this contribution, we have surveyed a number of computational methods including classical, quantum mechanical and modern data-driven statistical learning approaches for a theory-informed rational design of polymeric materials for dielectric and related applications. The examples discussed in this Perspective include computational strategies for the modification of existing polymers, as well as the discovery of novel polymer dielectrics and have guided successful experimental results. These success stories clearly demonstrate that carefully devised in silico design strategies, which combine computational methods over a range of time and length scales, are now capable of driving experimental studies toward materials of technological interest. Finally, we believe that the overarching ideas and screening strategies described in this paper are quite general and can easily be imported to address materials-related design challenges efficiently in other technological areas.

ACKNOWLEDGMENTS

This work was supported by a Multi-University Research Initiative (MURI) grant from the Office of Naval Research, under award number N00014-10-1-0944. Partial computational support of this research was provided by the National Science Foundation through XSEDE resources under Grant No. TG-DMR080058N and the Rensselaer Center for Biotechnology and Interdisciplinary Studies. Helpful discussions with Dr. Daniel Sinkovits, Mr. Arun Mannodi-Kanakkithodi and Dr. Huan Tran are also gratefully acknowledged.

REFERENCES

1. Sarjeant WJ, Zirnheld J, MacDougall FW. IEEE Trans Plasma Sci 1998;108: 1368e92.
2. Ho J, Jow R. Characterization of high temperature polymer thin films for power conditioning capacitors. Army Research Laboratories; Jul. 2009. ARL-TR-4880.
3. Zhu L, Wang Q. Macromolecules 2012;45:2937e54.
4. Chu B, Zhou X, Ren K, Neese B, Lin M, Wang Q, et al. Science 2006;313:334e6.
5. Wang Y, Zhou X, Chen Q, Chu BJ, Zhang QM. IEEE Trans Dielectr Electr Insul 2010;17:1036e42.
6. Wu S, Li WP, Li MR, Burlingame Q, Chen Q, Payzant A, et al. Adv Mater 2013;25:1734e8.
7. Yoon M-H, Yan H, Facchetti A, Marks TJ. J Am Chem Soc 2005;127:10388e95.
8. Bertolazzi S, Wünsche J, Cicoira F, Santato C. Appl Phys Lett 2011;99: 013301e3.
9. Roberts ME, Queralt N, Mannsfeld SCB, Reinecke BN, Knoll W, Bao Z. Chem Mater 2009;21:2292e9.
10. Facchetti A. Chem Mater 2011;23:733e58.
11. Dang MT, Hirsch L, Wantz G. Adv Mater 2011;23:3597e602.
12. Huo LJ, Zhang S, Guo X, Xu F, Li YF, Hou JH. Angew Chem 2011;123:9871e6.
13. Hohenberg P, Kohn W. Phys Rev 1964;136:B864e71.
14. Kohn W, Sham L. Phys Rev 1965;140:A1133e8.
15. Jones RO, Gunnarsson O. Rev Mod Phys 1989;61:689e746.
16. Baroni S, de Gironcoli S, Dal Corso A, Giannozzi P. Rev Mod Phys 2001;73: 515e62.

17. Sharma V, Pilania G, Rossetti GA, Slenes K, Ramprasad R. Phys Rev B 2013;87: 134109.
18. Wang CC, Ramprasad R. J Mater Sci 2011;46:90e3.
19. Wang CC, Pilania G, Ramprasad R. Phys Rev B 2013;87:035103.
20. Pilania G, Wang CC, Wu K, Sukumar N, Breneman C, Sotzing G, et al. J Chem Inf Model 2013;53(4):879e86.
21. Sun Y, Boggs SA, Ramprasad R. Appl Phys Lett 2012;101:132906.
22. Sun Y, Bealing C, Boggs SA, Ramprasad R. IEEE Electr Insul Mag 2013;29:8e15.
23. Ranjan V, Buongiorno-Nardelli M, Bernholc J. Phys Rev Lett 2012;108:087802.
24. Ranjan V, Yu L, Buongiorno-Nardelli M, Bernholc J. Phys Rev Lett 2007;99: 047801.
25. Huzayyin A, Boggs SA, Ramprasad R. IEEE Electr Insul Mag 2012;28:23e9.
26. Huzayyin A, Boggs SA, Ramprasad R. IEEE Trans Dielectr Electr Insul 2010;17: 920e5.
27. Huzayyin A, Boggs SA, Ramprasad R. IEEE Trans Dielectr Electr Insul 2010;17: 926e30.
28. Huzayyin A, Boggs SA, Ramprasad R. Annual report of the conference on electrical insulation and dielectric phenomena; 2010. pp. 1e4.
29. Jorgensen WL, Ulmschneider JP, Tirado-Rives J. J Phys Chem B 2004;108: 16264e70.
30. Wang JM, Wolf RM, Caldwell JW, Kollman PA, Case DA. J Comput Chem 2004;25:1157e74.
31. Vanommeslaeghe K, MacKerell Jr AD. J Chem Inf Model 2012;52:3144e54.
32. Vanommeslaeghe K, Prabhu Raman E, MacKerell Jr AD. J Chem Inf Model 2012;52:3155e68.
33. Sun H. J Phys Chem B 1998;102:7338e64.
34. Rigby D, Roe R. J Chem Phys 1987;87:7285e92.
35. Bitsanis I, Hadziioannou G. J Chem Phys 1990;92:3827e47.
36. Bennemann C, Paul W, Binder K. Phys Rev E 1998;57:843e51.
37. Moreno M, Casalegno M, Raos G, Meille SV, Po R. J Phys Chem B 2010;114(4): 1591e602.
38. Pilania G, Wang CC, Jiang X, Rajasekaran S, Ramprasad R. Sci Rep 2013;3:2810.
39. Sukumar N, Krein M, Luo Q, Breneman C. J Mater Sci 2012;47:7703e15.
40. Hughes LD, Palmer DS, Nigsch F, Mitchell JBO. J Chem Inf Model 2008;48(1): 220e32.
41. Hart GLW, Blum V, Walorski MJ, Zunger A. Nat Mater 2005;4:391e4.
42. Fischer CC, Tibbetts KJ, Morgan D, Ceder G. Nat Mater 2006;5:641e6.
43. Saad Y, Gao D, Ngo T, Bobbitt S, Chelikowsky JR, Andreoni W. Phys Rev B 2012;85:104104.
44. Rupp M, Tkatchenko A, Muller K, von Lilienfeld OA. Phys Rev Lett 2012;108: 058301.

45. Hautier G, Fisher CC, Jain A, Mueller T, Ceder G. Chem Mater 2010;22:3762e7.
46. Hansen K, Montavon G, Biegler F, Fazli S, Rupp M, Scheffler M, et al. J Chem Theory Comput 2013;9(8):3404e19.
47. Hammett LP. J Am Chem Soc 1937;59(1):96e103.
48. Hansch C, Muir RM, Fujita T, Maloney PP, Geiger F, Streich M. J Am Chem Soc 1963;85:2817e24.
49. Hansch C, Fujita T, editors. Classical and three-dimensional QSAR in agrochemistry. American Chemical Society symposium series. Washington: American Chemical Society; 1995.
50. Taft RW. J Am Chem Soc 1952;74:3120e8.
51. Katritzky AR, Lobanov VS, Karelson M. Chem Soc Rev 1995;24:279e87.
52. Rabuffi M, Picci G. IEEE Trans Plasma Sci 2002;30:1939e42.
53. Barshaw EJ, White J, Chait MJ, Cornette JB, Bustamante J, Folli F, et al. IEEE Trans Magn 2007;43:223e5.
54. Tortai JH, Bonifaci N, Denat A, Trassy CJ. Appl Phys 2005;97:053304(1)e 053304(9).
55. Yuan X, Matsuyama Y, Chung TCM. Macromolecules 2010;43:4011e4.
56. Chung TCM. Green Sustain Chem 2012;2:29e37.
57. Gupta S, Yuan X, Chung TCM, Kumar S, Weiss RA. Macromolecules 2013;46: 5455e63.
58. Wang CC, Pilania G, Ramprasad R, Agarwal M, Misra M, Kumar S, et al. Appl Phys Lett 2013;102:152901.
59. Grimme S. J Comput Chem 2006;27:1787e99.
60. Liu CS, Pilania G, Wang CC, Ramprasad R. J Phys Chem A 2012;116:9347e52.
61. Agarwal M, Misra M, Kumar S, Wang CC, Pilania G, Ramprasad R, et al. submitted.
62. Baldwin AF, Ma R, Wang CC, Ramprasad R, Sotzing GA. J Appl Polym Sci 2013;130:1276e80.
63. Lorenzini RG, Kline WM, Wang CC, Ramprasad R, Sotzing GA. Polymer 2013;54:3529e33.
64. Gonze X, Lee C. Phys Rev B 1997;55:10355.
65. Gonze X. Phys Rev B 1997;55:10337.
66. Oganov AR, Glass CW. J Chem Phys 2006;124:244704.
67. Lyakhov AO, Oganov AR, Valle M. Comput Phys Commun 2010;181: 1623e32.
68. Oganov AR, editor. Modern methods of crystal structure prediction. Berlin: Wiley-VCH; 2010, ISBN 978-3-527-40939-6; 2010.
69. Goedecker S. J Chem Phys 2004;120:9911.
70. Amsler M, Goedecker S. J Chem Phys 2010;133:224104.
71. Pannetier J, Bassas-Alsina J, Rodriguez-Carvajal J, Caignaer V. Nature 1990;346:343e5.

72. Kumar A, Wang CC, Ramprasad R. in preparation.
73. Robertson J. Rep Prog Phys 2006;69:327e96.
74. Robertson J. Solid State Electron 2005;49:283e93.
75. Zhu H, Ramprasad R. Phys Rev B 2011;83:081416.
76. Zhu H, Tang C, Fonseca RC, Ramprasad R. J Mater Sci 2012;47:7399e416.
77. Cao Y, Boggs SA. IEEE Trans Dielectr Electr Insul 2005;12:690e9.
78. Bealing CR, Ramprasad R. J Chem Phys 2013;139:174904.
79. Laurent C. In: Proc. of 1998 IEEE 6th Intl. Conf. on conduction and breakdown in solid dielectrics Sweden 1998.
80. von Hippel A. J Appl Phys 1937;8:815.
81. Zener CM. Proc R Soc 1934;145:523e9.
82. Frohlich H. J Appl Phys 1937;160:230.

Chapter 2

NANOPOLYMERS IMPROVE DELIVERY OF EXON SKIPPING OLIGONUCLEOTIDES AND CONCOMITANT DYSTROPHIN EXPRESSION IN SKELETAL MUSCLE OF MDX MICE

Jason H Williams[1], Rebecca C Schray[1], Shashank R Sirsi[2], and Gordon J Lutz[*1]

Address: [1]Drexel University College of Medicine, Department of Pharmacology and Physiology, Philadelphia, Pennsylvania 19102, USA and

[2]Drexel University, Department of Biomedical Engineering, Philadelphia, Pennsylvania 19104, USA

Email: Jason H Williams - jw85@drexel.edu; Rebecca C Schray - rcs46@drexel.edu; Shashank R Sirsi - srs43@drexel.edu;

Gordon J Lutz* - glutz@drexelmed.edu

[*]Corresponding author

ABSTRACT

Background

Exon skipping oligonucleotides (ESOs) of 2'O-Methyl (2'OMe) and morpholino chemistry have been shown to restore dystrophin expression in muscle fibers from the mdx mouse, and are currently being tested in phase I clinical trials for Duchenne Muscular Dystrophy (DMD). However, ESOs remain limited in their effectiveness because of an inadequate delivery profile. Synthetic cationic copolymers of poly(ethylene imine) (PEI) and poly(ethylene glycol) (PEG) are regarded as effective agents for enhanced delivery of nucleic acids in various applications.

Results

We examined whether PEG-PEI copolymers can facilitate ESO-mediated dystrophin expression after intramuscular injections into tibialis anterior (TA) muscles of mdx mice. We utilized a set of PEG-PEI copolymers containing 2 kDa PEI and either 550 Da or 5 kDa PEG, both of which bind 2'OMe ESOs with high affinity and form stable nanoparticulates with a relatively low surface charge. Three weekly intramuscular injections of 5 µg of ESO complexed with PEI2K-PEG550 copolymers resulted in about 500 dystrophin-positive fibers and about 12% of normal levels of dystrophin expression at 3 weeks after the initial injection, which is significantly greater than for injections of ESO alone, which are known to be almost completely ineffective. In an effort to enhance biocompatibility and cellular uptake, the PEI2K-PEG550 and PEI2K-PEG5K copolymers were functionalized by covalent conjugation with nanogold (NG) or adsorbtion of colloidal gold (CG), respectively. Surprisingly, using the same injection and dosing regimen, we found no significant difference in dystrophin expression by Western blot between the NG-PEI2K-PEG550, CG-PEI2K-PEG5K, and non-functionalized PEI2K-PEG550 copolymers.

Dose-response experiments using the CG-PEI2K-PEG5K copolymer with total ESO ranging from 3–60 µg yielded a maximum of about 15% dystrophin expression. Further improvements in dystrophin expression up to 20% of normal levels were found at 6 weeks after 10 twice-weekly injections of the NG-PEI2K-PEG550 copolymer complexed with 5 µg of ESO per injection. This injection and dosing regimen showed over 1000 dystrophinpositive fibers. H&E staining of all treated muscle groups revealed no overt signs of cytotoxicity.

Conclusion

We conclude that PEGylated PEI2K copolymers are efficient carriers for local delivery of 2′OMe ESOs and warrant further development as potential therapeutics for treatment of DMD.

Background

Steric block oligomers such as of 2′O-methyl (2′OMe) and phosphorodiamidate morpholino (PMO) oligonucleotides possess high affinity for their complementary premRNA targets and can modulate alternative splicing, correct aberrant splicing, and induce skipping or inclusion of specific exons (for reviews see [1,2]). These splice modulating oligonucleotides (SMOs) represent a powerful class of compounds with broad utility for basic and translational research and are poised to show rapid growth as pharmaceuticals. However, for many in vivo applications, SMOs administered alone show an inadequate delivery profile for reaching target cell nuclei, necessitating the use of carriers. Indeed, inefficient delivery of SMOs remains the foremost limitation to their usefulness as pharmaceuticals.

Duchenne muscular dystrophy (DMD) is a fatal x-linked disease caused by mutations in the gene encoding the 427 kDa membrane-associated cytoskeletal protein dystrophin, resulting in progressive body-wide muscle weakening and death usually in the early to mid third decade of life. DMD mutations are most often comprised of insertions and deletions that alter the dystrophin reading frame or encode premature stop codons [3]. These types of mutations result in production of truncated and nonfunctional dystrophin that is rapidly degraded.

Steric block exon skipping oligonucleotides (ESOs) have been shown in cultured mouse, canine, and human cells to cause skipping of targeted dystrophin exons, resulting in production of full-length dystrophin mRNA (minus only the skipped exons), and restoration of the reading frame [4-10]. The classical animal model for DMD, the mdx mouse, has a point mutation in dystrophin exon 23 that produces a premature stop codon. ESOs have been shown to promote skipping of exon 23 and concomitant dystrophin expression in skeletal muscles of mdx mice after both local and systemic delivery [8,11-17], and this technology in the context of therapy for DMD has been recently reviewed [18,19]. However, the efficiency of 2'OMe ESO delivery, and level of concomitant dystrophin expression in the mdx mouse remains relatively modest [11,13,16]. Despite the fact that ESOs without a carrier have progressed to phase I clinical trials, improved carriers must be developed before the ESO approach can be considered as a viable therapeutic for improving health and longevity in DMD patients.

The amine-rich cationic polymer poly(ethylene imine) (PEI) is a well-studied compound that is effective at condensing large plasmid DNA, enabling improved cellular uptake [20-25]. Although most often applied to plasmid delivery, recent studies have also documented PEIenhanced delivery of small nucleic acid agents such as oligonucleotides and siRNA [26-33]. The positive surface charge of PEI-nucleic acid polyplexes interacts with negatively charged elements on the cell membrane, stimulating non-specific receptor-mediated endocytotic uptake [20,21,24,25,34,35]. Once internalized, the 'proton sponge effect' enabled by PEI's buffering capacity induces the rupturing of the endosomal compartment due to osmotic lysis [34,36-38]. While the intracellular dynamics of PEI-nucleotide complexes once released from the endosome are unknown, oligonucleotides must dissociate from PEI to reach the nucleus, as the nuclear envelope is likely impermeable to PEI. Using dual-fluorescence tracking we recently showed PEG-PEI copolymers enhanced SMO delivery to myonuclei of cultured mdx myofibers, while the copolymers were mainly excluded from entering the nuclei [39].

The functionality and biocompatibility of PEI is greatly improved by incorporation of the nonionic linear polymer poly(ethylene glycol) (PEG) into PEG-PEI copolymers [23,24]. The macromolecular properties of PEG-PEIoligonucleotide polyplexes are greatly influenced by the molecular weight of PEI and nature of PEGylation, which in-turn effects transfection capacity [22,23,40]. We recently showed that PEG-PEI copolymers made of low MW PEI2K significantly improved delivery of ESOs to myofibers of mdx mice after intramuscular injections compared to high MW PEI25K copolymers [13]. We attributed the superior efficacy of the PEI2K-based copolymers to the low surface charge and high stability of the nanoparticles formed during complexation with ESOs [41]. Although intramuscular injection of these PEG-PEIESO polyplexes produced significantly more dystrophinpositive fibers than ESO alone, the level of dystrophin expression by Western blot reached only 2–5% of the normal level in TA muscles. Therefore, in this report we evaluated whether new injection regimens and further functionalization of PEG-PEI copolymers with gold nanoparticles might improve dystrophin expression. Our results show that repeat injections of small amounts of PEG-PEI-ESO are very effective at transducing dystrophin expression, reaching up to 20% of normal levels under the most favorable formulation and injection regimen.

Results

Induction of dystrophin expression following intramuscular injection of PEG-PEI-ESOs

The goal of this study was to determine an effective PEGPEI-ESO polyplex formulation and injection scheme for enhanced dystrophin expression after intramuscular injection into the TA muscle of 6–8 wk old male mdx mice. The ESO for all experiments was a 2'OMe oligonucleotide that was previously shown to produce specific skipping of mouse dystrophin exon 23, thereby removing a point mutation that encodes a premature stop codon [11,42]. Mice were given three weekly intramuscular injections of 5 μg of ESO complexed with either PEI2K(PEG550)10, NG-PEI2K(PEG550)10, or CGPEI2K(PEG5K)10 copolymers and were analyzed for dystrophin expression at one week after the third injection.

Immunohistochemistry of transverse sections showed that all three copolymer formulations produced substantially greater expression of dystrophin-positive fibers compared to muscles from un-injected mdx mice (Figure 1).

H&E staining showed morphological integrity to be well preserved, with no overt signs of muscle necrosis or cytotoxic damage (Figure 1). Quantitative evaluation of whole muscle cross-sections showed that injections with the PEI2K(PEG550)10-ESO formulation produced 594 ± 120 dystrophin-positive fibers (Figure 2), which corresponds to roughly 30% of the approximately 2000 fibers in the TA muscle [11,43].

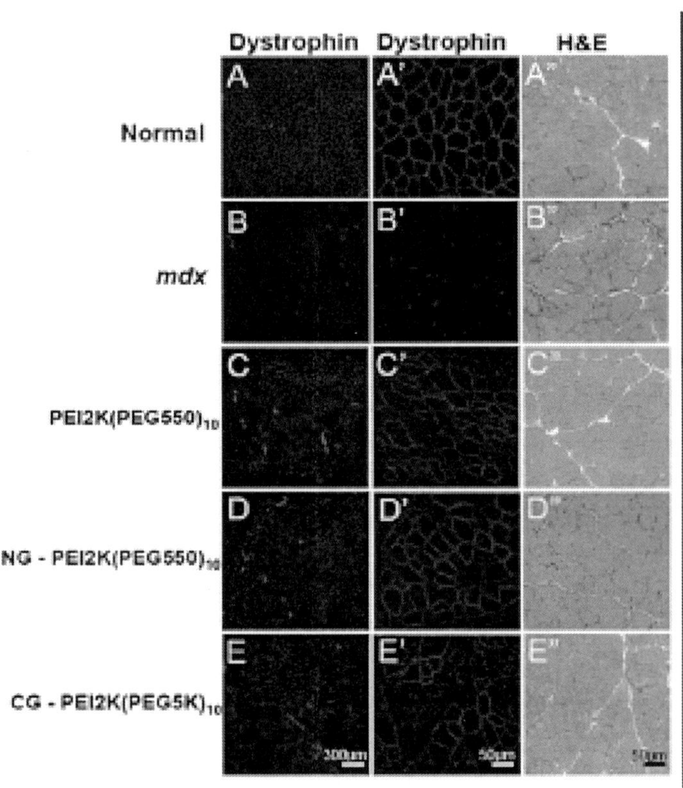

Figure 1. Dystrophin induction in limb musculature of mdx mice after intramuscular injections of 2′OMe ESO complexed with cationic nanopolymers. The TA muscles of mdx mice were given 3 weekly intramuscular injections of various PEG-PEI copolymers complexed with 5 μg of ESO and harvested 3 wks after the first injection. Dystrophin immunolabeling (Hoechst dye counterstained) at 2 different magnifications and H&E staining of serial transverse sections from TA muscles from the following groups: (A-A″) normal age-matched controls, (B-B″) mdx untreated, (C-C″) mdx injected with PEI2K(PEG550)10-ESO, (D-D″) mdx injected with NG-PEI2K(PEG550)10-ESO, and (E-E″) mdx injected with CG-PEI2K(PEG5K)10-ESO.

In accordance with previous studies showing improved biocompatibility and enhanced cellular uptake of goldconjugated nanoparticles [44,45], we hypothesized that conjugation or adsorption of nanogold (NG) or colloidal (CG) to PEG-PEI copolymers would improve the potency of these carriers. Surprisingly, injections with NGPEI2K(PEG550)10-ESO and CG-PEI2K(PEG5K)10-ESO resulted in only 380 ± 36 and 322 ± 24 dystrophin-positive fibers, respectively, both of which were significantly less than found with the unconjugated PEI2K(PEG550)10 copolymer (Figure 2A; $P < 0.05$).

To quantify dystrophin expression, Western blots were performed on whole muscle transverse sections directly serial to those on which fiber counts were obtained. Dystrophin expression following three weekly injections of the PEI2K(PEG550)10-ESO polyplex (5 μg ESO per injection) reached 11.2 +/- 2.4% of the level found in normal muscles (Figure 2B–C). As expected, dystrophin expression in control mdx muscles was not detectable, and was likely less than 1% of normal. The NG-PEI2K(PEG550)10-ESO and CG-PEI2K(PEG5K)10-ESO formulations resulted in 7.6 ± 1.0% and 9.8 ± 2.1% of the normal level of dystrophin expression, neither of which was statistically different than the unconjugated copolymer (Figure 2B–C). Although we did not perform injections of ESO without polymers in this study, we have shown in other studies that injections of 2′OMe ESO alone, under very similar dosing and injection regimens as used herein, produced very low numbers of dystrophin-positive fibers, and showed no detectable dystrophin on Western blots ([13] and Sirsi et al., manuscript submitted). Similarly, a previous study using this ESO delivered without a carrier showed no improvement in the number of dystrophin positive fibers over mdx control muscles [43]. The ineptitude of 2′OMe ESO delivery without a carrier in the mdx mouse has led to omitting ESO alone in other studies examining dystrophin induction following local and systemic delivery [11,16].

Dose-response profile of PEG-PEI-ESO polyplexes Using the same triple-injection regimen as above, the dose-response properties of CG-PEI2K(PEG5K)10-ESO polyplexes was evaluated over a range from 3–60 μg of total ESO. Dystrophin-positive fibers were detected at all dosages (Figure 3), but muscles injected with the smallest test dosage of ESO (1 μg × 3 injections) expressed significantly fewer dystrophin-positive fibers than doses of 15, 30, or 60 μg (Figure 4A; P

< 0.05). Western analysis showed that the highest level of dystrophin induction was achieved using 5 μg of ESO per injection, resulting in 14.2 ± 1.6% expression (Figure 4B–C), with apparent saturation at higher doses.

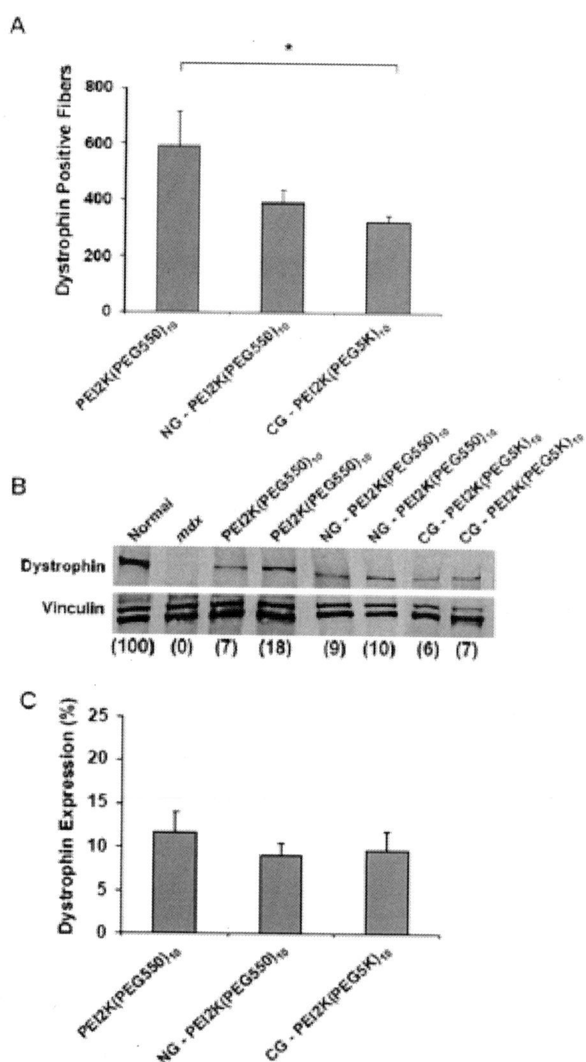

Figure 2. Quantitative analysis of dystrophin expression in TA muscles of mdx mice after intramuscular injections of PEG-PEI-ESO polyplexes. Muscles were analyzed following 3 weekly intramuscular injections of 5 μg of ESO

complexed with either PEI2K(PEG550)10, NG-PEI2K(PEG550)10, or CG-PEI2K(PEG5K)10 copolymers, and harvested 3 wks after the first injection. (A) Number of dystrophin-positive fibers for each treatment group was obtained from whole transverse sections that were immunolabeled for dystrophin. The number of dystrophin-positive fibers was significantly lower in muscles injected with NG-PEI2K(PEG550)10-ESO and CG-PEI2K(PEG5K)10-ESO polyplexes compared with the basic PEI2K(PEG550)10-ESO formulation ($P < 0.05$, $N = 4$ muscles per group). (B) Western blots show dystrophin expression in thick (60 μm) transverse cryosections taken directly adjacent to segments used for fiber counts in panel A. Images show blots of dystrophin (top) and vinculin (bottom) obtained from the same gel. All samples contained 25 μg of total protein, and dystrophin expression as a percent of normal is indicated in parentheses below each lane. (C) Quantitative western analysis of dystrophin expression as a percent of the level in age-matched normal mice for each of the three PEG-PEI-ESO polyplex formulations. No significant differences were observed between treatment groups ($P > 0.05$; $N = 4$ muscles per group).

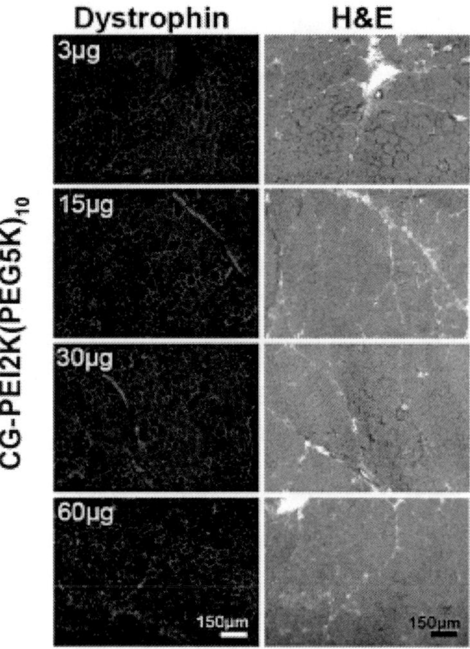

Figure 3. Dose-response profile of dystrophin expression after intramuscular injections of CG-PEI2K(PEG5K)10-ESO polyplexes. The TA muscles of mdx mice were given 3 weekly intramuscular injections of CG-PEI2K(PEG5K)10

copolymers complexed with 1, 5, 10, or 20 μg of ESO per injection (3, 15, 30, and 60 μg total) and harvested 3 wks after the first injection. Images show dystrophin immunolabeling (Hoechst dye counterstained) and H&E staining of serial transverse sections from TA muscles for each of the 4 polyplex doses.

Longer term repeat ESO injections increase dystrophin expression In an attempt to further improve dystrophin expression, we carried out longer term repeat injection experiments using the NG-PEI2K(PEG550)10-ESO polyplexes.

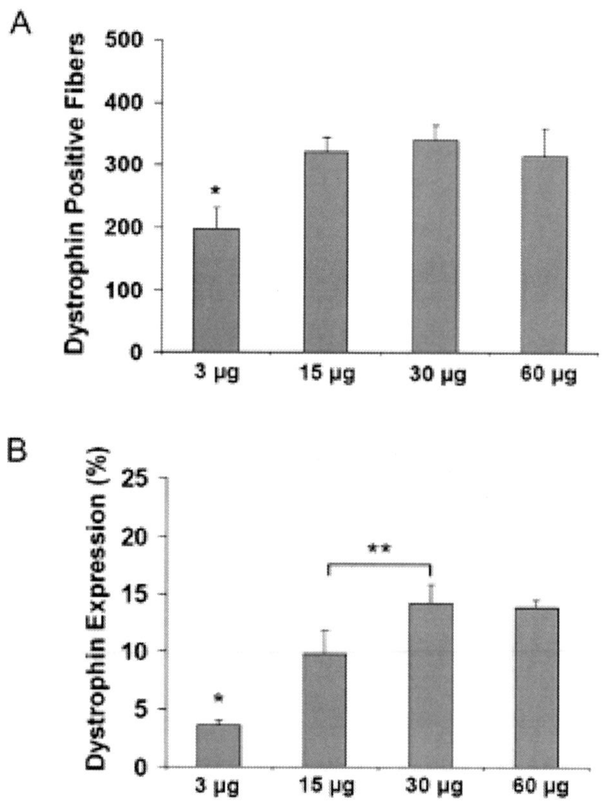

Figure 4. Quantitative analysis of dose-response profile of dystrophin expression after intramuscular injections of CG-PEI2K(PEG5K)10-ESO polyplexes. Muscles were analyzed following 3 weekly intramuscular injections of 1, 5, 10, or 20 μg of ESO (3, 15, 30, and 60 μg total) complexed with CG-PEI2K(PEG5K)10 copolymers, and harvested 3 wks after the first injection. (A) Number of dystrophin-positive fibers for each dosage was obtained from

whole transverse sections that were immunolabeled for dystrophin. Number of dystrophin-positive fibers was significantly lower for the 1 µg injections than for all other doses ($P < 0.05$, $N = 4$ muscles per group). (B) Quantification of dystrophin expression as a percentage of age-matched normal mice for each of the 4 polyplex dosages based on densitometry of western blots (not shown). Samples were prepared from thick (60 µm) transverse cryosections taken adjacent to segments used for the fiber counts. Each increase in dosage resulted in significantly greater level of dystrophin expression, except between the 30 µg and 60 µg dosages which reached a plateau.

Mice were given 10 consecutive intramuscular injections of NG-PEI2K(PEG550)10-ESO polyplexes (1 or 5 µg of ESO per injection), with 4 days between injections, and were harvested at 6 weeks after the first injection. Immunohistochemistry of whole transverse sections showed that this protocol resulted in marked improvement in ESO delivery to myofibers, producing extensive regions of intensely labeled dystrophin-positive fibers (Figure 5).

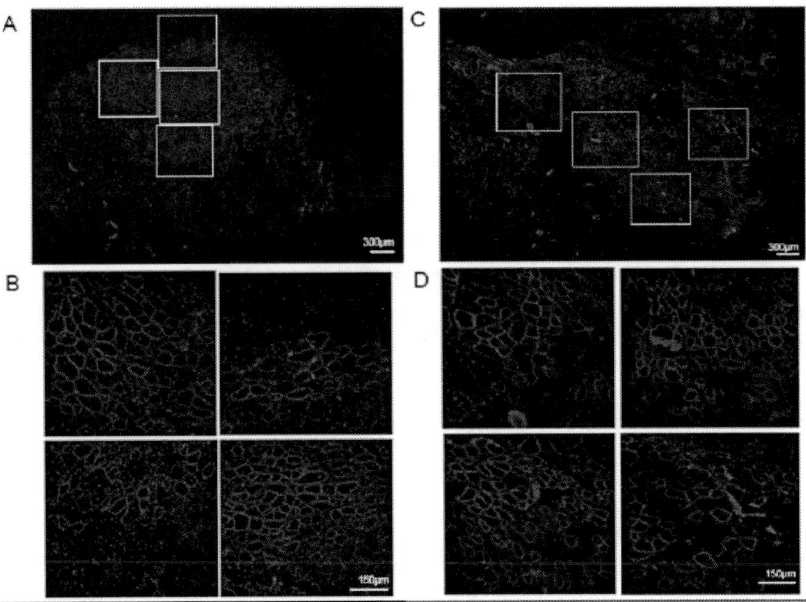

Figure 5. A 6 week, repeat intramuscular injection, regimen of PEG-PEI-ESO polyplexes produced widespread manifestation of dystrophin-positive fibers. Dystrophin immunolabeling of whole transverse sections of TA muscles of mdx mice is shown after 10 twice-weekly intramuscular injections of either

1 μg (A-B) or 5 μg (C-D) of ESO complexed with the NGPEI2K(PEG550)10 copolymer. Muscles were harvested at 6 wks after the first of the 10 injections. High magnification views of four individual regions of each muscle section (boxed) are shown.

On average, muscles injected with the 1 and 5 μg of ESO per injection (10 and 50 μg total ESO over 6 weeks) contained 832 ± 167 and 1225 ± 343 dystrophin-positive fibers, respectively (Figure 6B). The number of dystrophin-positive fib ers after delivery of 10 μg total ESO for 6 weeks was 2.2-fold greater than found after 15 μg of ESO for 3 weeks using the same NG-PEI2K(PEG550)10 copolymer. H&E staining confirmed the increased dystrophin expression following repeated injections was not associated with any overt signs of cytotoxicity (Figure 6A).

Western blots further established the potency of the 10-repeat injection regimen using the NG-PEI2K(PEG550)10-ESO polyplexes (Figure 6C–D). On average, muscles injected with 50 μg of ESO expressed 19.1 ± 1.8% of the normal amount of dystrophin, a level which was previously shown to be sufficient for improvement in mdx muscle mechanical properties [43,46,47]. The dystrophin expression after 50 μg of ESO was significantly greater than the 8.1 ± 2.8% produced with 10 μg of ESO ($P < 0.05$). However, there was exceptionally high variability in dystrophin expression within the 10 μg group as the lowest and highest values ranged from 4 to 23%. The peak level of 23% dystrophin expression after only 10 μg ESO injected over 6 weeks demonstrates the effectiveness of the nanopolymer-ESO formulation coupled with a low dose-high frequency delivery schedule.

Expression of nNOS in mdx muscles treated with PEG-PEIESOs

Muscles from polyplex-treated mdx mice were analyzed for nNOS expression as evidence for expression of functional dystrophin with an intact N-terminal binding domain. It was previously shown that nNOS expression is absent from muscles of mdx mice [48] and may be upregulated following exon-skipping restoration of dystrophin expression [11]. Therefore, we asked whether the current strategy for restoring dystrophin expression increased the level of nNOS protein localized at the myofiber membrane, where it serves to maintain normal cellular function in close association with dystrophin. Serial sections immunolabeled for dystrophin and nNOS showed a very tight correlation between dystrophin- and nNOS-positive fibers (Figure 7). Specifically, membrane-associated nNOS appeared to be expressed only in the mdx muscle cells that also showed enhanced dystrophin expression, indicating that ESO-expressed dystrophin was functionally intact. As expected, nNOS expression was absent from mdx un-injected control muscles.

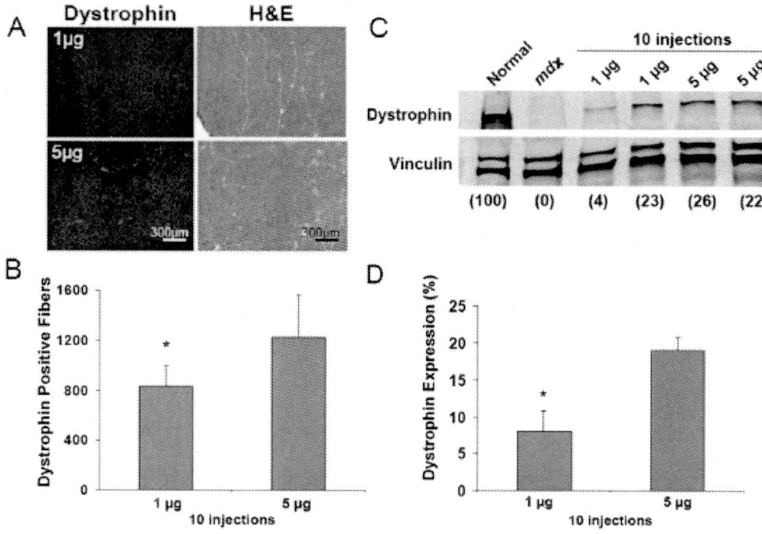

Figure 6. NGQuantitative analysis of dystrophin expression following a 6 week, repeat intramuscular injection regimen of NG-PEI2K(PEG550)10 polyplexes. Tissues were analyzed following 10 twice-weekly intramuscular injections of either 1 or 5 μg of ESO complexed with NG-PEI2K(PEG550)10 copolymers, and harvested 6 wks after the first injection. (A) Dystrophin immunolabeling (Hoechst dye counterstained) and H&E staining of serial transverse sections from TA muscles from the 10 μg (1 μg × 10 injections) and 50 μg (5 μg × 10 injections) groups. (B) Number of dystrophin-positive fibers, obtained from whole transverse sections immunolabeled for dystrophin, was significantly lower in muscles injected with 10 μg ESO compared with the 50 μg group ($P < 0.001$, $N = 4$ muscles per group). (C) Western blots showing dystrophin expression in thick (60 μm) transverse cryosections taken adjacent to segments used for the fiber counts in panel B. Images show blots of dystrophin (top) and vinculin (bottom) obtained from the same gel. All samples contained 25 μg of total protein, and dystrophin expression as a percent of normal is indi-

cated in parentheses below each lane. (D) Dystrophin expression determined from Western blots reached 20% of normal levels in muscles injected with NG-PEI2K(PEG550)10-ESO containing 50 μg ESO, which was significantly greater than the 10 μg group ($P < 0.001$, $N = 4$ muscles per group).

Discussion

The growing opinion that ESOs are on the path to becoming a viable therapeutic option for DMD is well supported by cell culture and animal data showing specific skipping of various targeted exons and resultant induction of nearly full-length dystrophin [4,6,7,9,11-16,49,50]. However, there is vigorous debate as to which ESO chemistry may work best, and what type of carrier compound will provide adequate delivery to body musculature. ESOs of phosphorothioate 2′OMe and PMO chemistry have been the best studied in animal models for DMD and both are currently being tested in Phase I clinical safety trials [51].

Both 2′OMe and PMO ESOs function by sterically blocking pre-mRNA target sequences, and it is thought that they may be used interchangeably once an optimal sequence (for a single chemistry) has been empirically determined [52]. However, fundamental differences in 2′OMe and PMO backbone chemistry preclude the use of a universal carrier for efficient delivery. Specifically, PMOs are synthetic compounds that are extraordinarily resistant to chemical degradation, but they are also charge neutral which limits cell surface interactions and cellular uptake. The non-degradable nature of PMOs also raises concerns over their safety after extended applications In contrast, 2′OMe ESOs are anionic RNA, and despite improved stability due to their phosphorothioate backbone, remain somewhat susceptible to degradation, while the negative charge hinders biodistribution and cellular uptake.

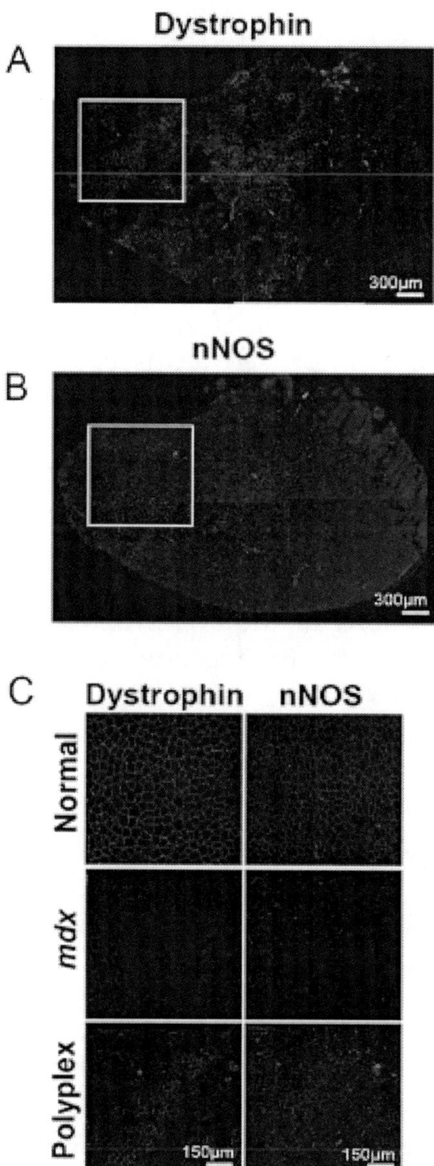

Figure 7. nNOS expression in mdx mice after treatment with PEG-PEI-ESO. Membrane-associated nNOS expression is upregulated in PEG-PEI-ESO treated mdx muscles in regions highly positive for dystrophin expression. TA muscles were given 10 twice-weekly intramuscular injections of 5 μg of ESO

complexed with the NG-PEI2K(PEG550)10 copolymer, and harvested 6 wks after the first injection. (A-B) Images of serial whole transverse sections immunolabeled for dystrophin (A) and nNOS (B) show the correlation in ESO-mediated upregulation of these two membrane-associated proteins. (C) Higher magnification images of specific regions from panels A and B show clearly the concomitant (co-localized) upregulation of both dystrophin and nNOS in the same set of fibers of the polyplex-injected muscles. As expected nNOS expression in untreated mdx controls was negligible.

In this study we showed that PEG-PEI copolymers formulated with low MW PEI2K function as effective carriers for delivery of 2′OMe ESOs to myofibers of mdx mice after intramuscular injections, resulting in improved levels of dystrophin expression. Specifically, three weekly intramuscular injections of only 5 μg of ESO complexed with the PEI2K(PEG550)10 copolymer resulted in about 600 dystrophin-positive fibers and about 11% of the normal level of dystrophin expression at 3 weeks after the initial injection. Still higher levels of dystrophin expression were achieved using a twice-weekly injection regimen extended out to 6 weeks. Specifically, 10 consecutive injections of the NG-PEI2K(PEG550)10 copolymer complexed with 5 μg of ESO produced over 1200 dystrophin positive fibers and 20% of normal levels of dystrophin expression. In regions with the most highly transfected fibers, we observed a concomitant increase in membrane-associated nNOS, specifically in dystrophin-positive fibers, in agreement with previous reports using this ESO [11]. Our lab, as well as other, have demonstrated that intramuscular injections of 2′OMe ESO alone, using very similar conditions as used herein, produced very few dystrophin-positive fibers and negligible levels of dystrophin on western blots [13,43]. In addition, we previously showed that single injections of the PEI2K(PEG550)10 copolymer complexed with 20 μg of ESO produced about 460 dystrophin-positive fibers at 3 weeks after the injection, but western blots showed dystrophin expression was only 2–5% of normal levels [13]. Taken together, the current results suggest that PEG-PEI copolymers enhance dystrophin expression and that repeat injections are more effective at transfecting a greater number of muscle fibers than individual injections containing about the same amount of ESO.

Although the dystrophin expression levels shown in this report demonstrate the utility of PEG-PEI nanopolymers for delivery of

ESOs, these compounds appear to be somewhat less effective than PMOs. Specifically, Alter et al [14] recently showed that single intramuscular injections of 10 μg of PMO resulted in up to 60% of the normal level of dystrophin expression, although this report appeared to provide only an estimate of efficacy and lacked statistical validation. Other recent studies of PMOs with conjugated peptide cell-targeting moieties showed impressive numbers of dystrophin-positive fibers, but did not provide a thorough evaluation of dystrophin expression by western blots [12].

The PEI2K(PEG550)10 and PEI2K(PEG5K)10 copolymers utilized in this study were previously shown to form exceptionally stable complexes when mixed with negatively charged ESO and the surface charge of the resultant nanoparticulates was relatively low [41]. We propose that the high stability and low surface charge of these polyplexes are two salient features that make them better suited for in vivo delivery of ESOs than high MW PEI25Kbased copolymers. Specifically, the low polyplex surface charge favors biodistribution and reduces cytotoxicity, while the high stability allows the polyplex to remain associated during extracellular to intracellular trafficking.

Gold nanoparticles such as NG and CG have been shown to improve biocompatibility and enhance cellular uptake of various types of cargo in drug delivery applications [44,45,53,54]. In particular, NG conjugated to low MW PEI2K showed at least an order of magnitude greater efficiency than PEI25K and was 12 times more potent than unmodified PEI2K for delivery of plasmid DNA in cell culture [44]. Unexpectedly, our results showed that neither covalent conjugation of NG or electrostatic surface coating with CG of the PEG-PEI copolymers improved ESO delivery. Polyplex stability assays in PBS showed CG and NG caused only moderate weakening of polyplex stability (data not shown), making this an unlikely explanation for the lack of improved delivery. A possible explanation for the lack of improvement is that the CG coating was not stable enough to adhere to the copolymer during delivery.

On the other hand, NG was covalently conjugated to PEGPEI, and we postulate that its ineffectiveness was more likely due to the 1:10 NG to PEI2K ratio, which may have been too low to improve functionality.

The dystrophin expression achieved in the present study was accomplished with PEG-PEI carriers that did not appear to elicit any overt signs of cytotoxicity. This is in contrast to previous studies using non-PEGylated PEI25K as a carrier of 2′OMe ESOs, which was ineffective and resulted in significant damage following only a very limited number of injections [55,56]. Based on these results, it was concluded that cationic polymers are unsuitable for in vivo delivery of AO in skeletal muscle [11,57]. However, the PEI-nucleotide particles used in these previous studies undoubtedly had very high positive surface charges, because they did not contain PEG, which is known to provide steric shielding of the PEI surface charge. In addition, dispersion of the highly-charged particles after intramuscular injection is probably severely hindered by charge interactions between the PEI and negatively-charged elements within the extracellular environment. Therefore, it is not surprising that these previous muscle transfection studies with non-PEGylated PEI produced unsatisfactory results. We suggest that the combination of a low MW PEI and extensive PEGylation used presently provided a favorable formulation which was both effective and non-toxic.

However, the lack of cytotoxicity observed does not preclude the possibility that some damage to muscle occurs immediately after injection, resulting in some level of degeneration-regeneration. This process may underlie to some extent the high number of dystrophin-positive fibers observed in our 6 week (10 injection) trials. Although not systematically evaluated, we have observed that shortterm mechanical damage occurs in mdx muscles after intramuscular injections of various solutions (even saline) that does not occur in normal muscle. Because of the lack of dystrophin, mdx muscles are more susceptible to mechanical damage than normal muscle. This effect may be exacerbated to some extent by cationic particles, or for that matter, any type of carrier compound.

We suggest that the major limitation of the carrier-ESO formulations described in this report was inadequate carrier functionality, and not a lack of intrinsic potency of the ESO. The ESO used in the present study (designated in the literature as M23D(+02–18)) has been shown in vitro to predominantly produce skipping of exon 23, although some exon 22–23 double skipping does occur [42]. Moreover, in this study we showed that under the most effective condition, about 50% of fibers were dystrophinpositive, resulting in about 20% of normal dystrophin expression. This indicates that on average dystrophin-positive fibers contained about 40% of the normal level of dystrophin. A similar calculation based on data reported by Lu et al [11], and our recent study with TAT-conjugated copolymers (Sirsi et al., manuscript submitted) suggests that dystrophin per transfected fiber may reach 75% of normal levels. Thus, the main limitation with cationic carriers seems to be their poor diffusional distribution, as indicated by large regions in muscles with no apparent transfection. Thus, further improvements in carrier functionality will likely be required to enable their usage in a clinical setting for DMD. Our group recently showed that conjugation of multiple HIV-TAT epitopes to PEI2K(PEG5K)10 copolymers greatly improved ESO delivery, using a similar dosing and intramuscular injection regimen as reported here, resulting in up to 30% dystrophin expression (Sirsi et al., manuscript submitted).

Various other types of cell targeting ligands, cell penetrating peptides, or fusogenic peptides may also be conjugated to PEI to improve functionality. Importantly, this type of peptide-PEI conjugate can likely be formulated for improved systemic delivery, which will be required to achieve meaningful therapeutic benefit. PMOs have already been shown to have limited efficacy after systemic delivery. For example, intraperitoneal injections into neonatal mdx mice of 5 mg/kg/week with PMO-peptide produced widespread dystrophin in diaphragm muscle, with low levels of expression observed in limb muscles [12,58].

CONCLUSION

In this report, we show that PEGylated amine-rich branched cationic nanopolymers comprised of low MW PEI2K are effective carriers for delivery of ESOs to myofibers of mdx mice after intramuscular injections. Our results indicate that high frequency, low-dosage, long-term injection regimens using these carrier-ESO compounds provide the most favorable outcome in terms of dystrophin expression. While other studies have shown improvements utilizing gold nanoparticles, we showed no enhancement when conjugated to the current PEG-PEI formulations. Despite the lack of improvement using these novel conjugations, PEGylated PEI2K copolymers remain as one of the most efficient carriers for local delivery of 2′OMe ESOs and warrant further development as potential therapeutics for treatment of DMD.

METHODS

Animals

Male mdx mice (C57BL/10ScSn-Dmdmdx/J) and agematched 6–9 wk old normal mice (C57BL/10SnJ) were obtained from Jackson Laboratories (Bar Harbor, ME). All animals were housed according to NIH and University guidelines (Drexel University College of Medicine, ULAR facility, Philadelphia, PA).

Nanopolymer synthesis

The synthesis and physiochemical characterization of the PEI2K(PEG550)10 and PEI2K(PEG5K)10 copolymers were previously described [41,59]. Copolymers are designated using a nomenclature where the subscript indicates the number of PEG chains grafted per molecule of PEI. For example, PEI2K(PEG550)10 indicates 10 PEG chains of 550 daltons grafted to each 2 kDa PEI molecule. Nanogold (NG) particles were conjugated to PEI primary amine groups on PEI2K(PEG550)10 coploymers using the Sulfo N- Hydroxy-Succinimido Nanogold labeling reagent (Nanoprobes, Yaphank, NY). In this reaction, 75 nmols of PEI2K(PEG550)10 was mixed with 6 nmols of NHS-nanogold in 780 µl of sterile water (pH = 8.0). The solution was incubated for 24 hours on ice, frozen, and subsequently freeze-dried and stored at -20°C. Adsorption of colloidal gold (CG) to PEI2K(PEG5K)10 was performed by mixing 300 µl of 5 nM CG particles (Sigma-Aldrich) with 30 mg of PEI2K(PEG5K)10 (in 1 ml DI H20) and incubating at 4°C overnight. The polymer solution was subsequently freeze-dried and stored at -20°C. The NG and CG labeled copolymers are designated as NG-PEI2K(PEG550)10 and CG-PEI2K(PEG5K)10, respectively.

Preparation of PEG-PEI-ESO polyplexes

The ESO used in all experiments was a 20-mer oligoribonucleotide

(5′-GGCCAAACCUCGGCUUACCU-3′) previously shown to cause skipping of dystrophin exon 23 in mdx mice [11,16,42]. During synthesis (Trilink, San Diego, CA) each base was phosphorothioated and contained a methoxy group at the 2′ carbon. PEG-PEI-ESO polyplexes were prepared at a nitrogen to phosphate ratio of 5 (N:P = 5); where N represents moles of amine on PEI, and P represents moles of phosphate on ESO. Polyplexes were formed by the addition of the PEG-PEI copolymer solution to the ESO solution (in sterile saline). Polyplex solutions were vortexed briefly, sonicated for 30 min using a bath sonicator, incubated on ice for 30 min, and used immediately.

Intramuscular injections of PEG-PEI-ESO polyplexes

Male mdx mice (6–9 weeks of age) were anesthetized with ketamine/xylazine and shaved for visualization of hindlimb muscles. A 15 µl volume of PEG-PEI-ESO polyplex solution at various concentrations was injected bi-laterally into the mid-belly portion of TA muscles using a 31 gauge insulin syringe. After recovery from anesthesia, mice were returned to normal cage activity.

We used both a 3 wk and 6 wk polyplex injection schedule as follows. For the 3 wk groups, mice were injected on days 0, 7, and 14; and were harvested on day 21. In one series of experiments muscles were injected with 5 µg of ESO (15 µg total) complexed with either the PEI2K(PEG550)10, NG-PEI2K(PEG550)10, or CGPEI2K(PEG5K)10 copolymers. For dose-response analysis of a single polyplex, muscles were injected with 1, 5, 10, or 20 µg of ESO (3, 15, 30, and 60 µg total) complexed with the CG-PEI2K(PEG5K)10 copolymer. For the 6 wk groups, mice were injected on days 0, 4, 8, 12, 16, 20, 24, 28, 32, and 36; and were harvested on day 42. For this experiment, muscles were injected with either 1 or 5 µg of ESO per injection (10 and 50 µg total) complexed with the NG-PEI2K(PEG550)10 copolymer. For all groups, 2 mice (N = 4 muscles) were analyzed.

Muscle harvest, immunohistochemistry, histology, and fiber counts

At designated time points mice were killed and TA muscles were removed, pinned to parafilm-covered cork, snap frozen in liquid

N2-cooled isopentane, and stored at-80°C. Control muscles were harvested from uninjected age-matched mdx and normal mice. Transverse frozen sections (10 µm) were cut from the belly of each TA muscle using a cryostat (Leica CM 3050 S, Bannockburn, IL) and melted onto slides for immunohistochemistry and histochemistry.

Thick (60 µm) transverse sections, immediately adjacent to the thin sections were cut in the cryostat and placed in 1.5 mL centrifuge tubes on dry ice, and stored at -80°C for subsequent western analysis.

For immunolabeling, muscle sections were blocked with 10% normal goat serum in 1% BSA/PBS for 1 h and then incubated for 1 h in rabbit polyclonal anti-dystrophin (1:200; Abcam, Cambridge, MA) or anti-nNOS (1:125; Invitrogen, Carlsbad, CA). The secondary antibody was Cy3-Anti-Rabbit IgG (1:500; Jackson Immunoresearch, West Grove, PA). Slides were coverslipped with Vectashield mounting medium with DAPI (Vector Laboratories, Burlingame, CA) and imaged (4/10/20× objective; Olympus, AX70, Melville, NY). Composite images of entire transverse sections were constructed from overlapping low magnification images using Adobe Photoshop (Adobe, San Jose, CA). Fiber counts of dystrophin-positive fibers were obtained using the cell counter function of ImageJ software (rsb.info.nih.gov/ij/plugins/cell-counter. html).

Western analysis of dystrophin expression

Sections of frozen TA muscles (60 µm) were extracted in 1.5 ml centrifuge tubes by pipetting up and down in 50 µl of protein extraction buffer containing 125 mM Tris (pH 6.8), 4% SDS, 2 M Urea, 5% 2-mercaptoethanol, 10% glycerol, 5 µl of protease inhibitor cocktail (Sigma, St. Louis, MO) and protease inhibitors calpeptin (100 nM; Calbiochem, San Diego, CA) and calpain inhibitor I (25 µM; Calbiochem, San Diego, CA). The extract was incubated on ice (15 min), boiled (5 min), and centrifuged (4000 × g for 5 min), and the supernatant was transferred to a clean tube. Protein concentration was measured in extracts using the Coomassie assay (Pierce, Rockford, IL) and an equal volume of SDS-PAGE sample buffer (125 mM Tris (pH 6.8), 4% SDS, 5% 2-mercaptoethanol, 10% glycerol and 0.05% bromophenol blue) was added to the extract. Samples containing 25 µg of total protein were loaded onto pre-cast SDS-PAGE gels (3%

stacking: 7.5% resolving; Bio-Rad, Hercules, CA) and run at 150 V for 75 min. Gels were transferred to nitrocellulose at 30 V for 16 h and membranes were stained with Ponceau S (Sigma, St. Louis, MO) to visualize proper transfer and even loading. Membranes were cut to allow separate immunoblotting of dystrophin and vinculin and blocked in Odyssey blocking buffer (LI-COR Biosciences, Lincoln, NE) for 1 h. Membranes were incubated for 1 h in mouse monoclonal antidystrophin (MANDYS8; Sigma, St. Louis, MO) and antivinculin (VIN1; Sigma, St. Louis, MO) at dilutions of 1:400 and 1:2000, respectively. Donkey anti-mouse IRDye 800 CW secondary antibody (LI-COR Biosciences, Lincoln, NE) was applied for 1 h and membranes were scanned on an Odyssey Infrared Imaging System following multiple TBS-T/TBS washes. Odyssey imaging software was used for densitometry and the integrated intensity of sample bands was used for calculating the percentage of dystrophin expression as compared to the normal muscles. For each muscle, 2–3 separate 60 µm sections were extracted and used in western analysis.

Statistical Analysis

All data are reported as mean values ± SEM. Statistical differences between treatment groups were evaluated by ANOVA (Statview; SAS Institute, Cary, NC).

List of abbreviations

ESO, Exon skipping oligonucleotide; 2'OMe, 2'-OMethyl; PMO, Phosphorodiamidate Morpholino Oligonucleotide; CPP, Cell Penetrating Peptide; DMD, Duchenne Muscular Dystrophy; PEI, Poly(ethylene imine); PEG, Poly(ethylene glycol); N:P, ratio of PEI nitrogen to ESO phosphate; TA, Tibialis Anterior; NG, nanogold; CG, Colloidal gold.

Competing interests

The author(s) declare that they have no competing interests.

AUTHORS' CONTRIBUTIONS

JHW carried out western and immunoflourescence analysis, assisted with harvesting of muscle tissue, participated in the design of the study and drafted the manuscript. RCS maintained animal colonies, carried out injections and harvesting of tissues and participated in design of the study. SRS carried out chemical synthesis and functionalization of copolymers and contributed to the conception of the study. GJL conceived the study, drafted the final manuscript and participated in all stages of the work. All authors read and approved the final manuscript.

ACKNOWLEDGEMENTS

We thank Michelle Erney for assistance with the immunofluorescence and Nilo Pebdani and Elisa McDaniel for assistance with the Western blotting. This work was supported by a grant from the Muscular Dystrophy Association (USA).

REFERENCES

1. Kole R, Williams T, Cohen L: RNA modulation, repair and remodeling by splice switching oligonucleotides. Acta Biochim Pol 2004, 51:373-378.
2. Garcia-Blanco MA, Baraniak AP, Lasda EL: Alternative splicing in disease and therapy. Nat Biotechnol 2004, 22:535-546.
3. Aartsma-Rus A, van Deutekom JC, Fokkema IF, van Ommen GJ, den Dunnen JT: Entries in the Leiden Duchenne muscular dystrophy mutation database: an overview of mutation types and paradoxical cases that confirm the reading-frame rule. Muscle Nerve 2006, 34:135-144.
4. Aartsma-Rus A, Janson AA, Kaman WE, Bremmer-Bout M, den Dunnen JT, Baas F, van Ommen GJ, van Deutekom JC: Therapeutic antisense-induced exon skipping in cultured muscle cells from six different DMD patients. Hum Mol Genet 2003, 12:907-914.
5. Aartsma-Rus A, Kaman WE, Weij R, den Dunnen JT, van Ommen GJ, van Deutekom JC: Exploring the frontiers of therapeutic exon skipping for duchenne muscular dystrophy by double targeting within one or multiple exons. Mol Ther 2006, 14:401-407.
6. McClorey G, Moulton HM, Iversen PL, Fletcher S, Wilton SD: Antisense oligonucleotide-induced exon skipping restores dystrophin expression in vitro in a canine model of DMD. Gene Ther 2006, 13:1373-1381.

7. Gebski BL, Mann CJ, Fletcher S, Wilton SD: Morpholino antisense oligonucleotide induced dystrophin exon 23 skipping in mdx mouse muscle. Hum Mol Genet 2003, 12:1801-1811.

8. Mann CJ, Honeyman K, Cheng AJ, Ly T, Lloyd F, Fletcher S, Morgan JE, Partridge TA, Wilton SD: Antisense-induced exon skipping and synthesis of dystrophin in the mdx mouse. Proc Natl Acad Sci U S A 2001, 98:42-47.

9. Aartsma-Rus A, Janson AA, Kaman WE, Bremmer-Bout M, van Ommen GJ, den Dunnen JT, van Deutekom JC: Antisense-induced multiexon skipping for Duchenne muscular dystrophy makes more sense. Am J Hum Genet 2004, 74:83-92.

10. Aartsma-Rus A, Janson AA, van Ommen GJ, van Deutekom JC: Antisense-induced exon skipping for duplications in Duchenne muscular dystrophy. BMC Med Genet 2007, 8:43.

11. Lu QL, Mann CJ, Lou F, Bou-Gharios G, Morris GE, Xue SA, Fletcher S, Partridge TA, Wilton SD: Functional amounts of dystrophin produced by skipping the mutated exon in the mdx dystrophic mouse. Nat Med 2003, 9:1009-1014.

12. Fletcher S, Honeyman K, Fall AM, Harding PL, Johnsen RD, Steinhaus JP, Moulton HM, Iversen PL, Wilton SD: Morpholino Oligomer-Mediated Exon Skipping Averts the Onset of Dystrophic Pathology in the mdx Mouse. Mol Ther 2007, 15(9):1587-1592.

13. Williams JH, Sirsi SR, Latta D, Lutz GJ: Induction of dystrophin expression by exon skipping in mdx mice following intramuscular injection of antisense oligonucleotides complexed with PEG-PEI copolymers. Mol Ther 2006, 14(1):88-96.

14. Alter J, Lou F, Rabinowitz A, Yin H, Rosenfeld J, Wilton SD, Partridge TA, Lu QL: Systemic delivery of morpholino oligonucleotide restores dystrophin expression bodywide and improves dystrophic pathology. Nat Med 2006, 12:175-177.

15. Fletcher S, Honeyman K, Fall AM, Harding PL, Johnsen RD, Wilton SD: Dystrophin expression in the mdx mouse after localised and systemic administration of a morpholino antisense oligonucleotide. J Gene Med 2005, 8(2):207-216.

16. Lu QL, Rabinowitz A, Chen YC, Yokota T, Yin H, Alter J, Jadoon A, Bou-Gharios G, Partridge T: From the Cover: Systemic delivery of antisense oligoribonucleotide restores dystrophin expression in body-wide skeletal muscles. Proc Natl Acad Sci U S A 2005, 102:198-203.

17. Wilton SD, Lloyd F, Carville K, Fletcher S, Honeyman K, Agrawal S, Kole R: Specific removal of the nonsense mutation from the mdx dystrophin mRNA using antisense oligonucleotides. Neuromuscul Disord 1999, 9:330-338.

18. Aartsma-Rus A, van Ommen GJ: Antisense-mediated exon skipping: A versatile tool with therapeutic and research applications. RNA 2007, 13:1609-1624. 19. Wilton SD, Fletcher S: Modification of pre-mRNA processing: application to dystrophin expression. Curr Opin Mol Ther 2006, 8:130-135.

19. Bieber T, Meissner W, Kostin S, Niemann A, Elsasser HP: Intracellular route and transcriptional competence of polyethylenimine-DNA complexes. J Control Release 2002, 82:441-454.
20. Suh J, Wirtz D, Hanes J: Efficient active transport of gene nanocarriers to the cell nucleus. Proc Natl Acad Sci U S A 2003, 100:3878-3882.
21. Petersen H, Fechner PM, Martin AL, Kunath K, Stolnik S, Roberts CJ, Fischer D, Davies MC, Kissel T: Polyethylenimine-graftpoly(ethylene glycol) copolymers: influence of copolymer block structure on DNA complexation and biological activities as gene delivery system. Bioconjug Chem 2002, 13:845-854.
22. Petersen H, Fechner PM, Fischer D, Kissel T: Synthesis, Characterization, and Biocompatibility of Polyethylenimine-graftpoly(ethylene glycol) Block Copolymers. Macromolecules 2002, 35:6867-6874.
23. A: Gene transfer with modified polyethylenimines. J Gene Med 2004, 6 Suppl 1:S3-10.
24. Thomas M, Klibanov AM: Non-viral gene therapy: polycationmediated DNA delivery. Appl Microbiol Biotechnol 2003, 62:27-34.
25. Vinogradov S, Batrakova E, Li S, Kabanov A: Polyion complex micelles with protein-modified corona for receptor-mediated delivery of oligonucleotides into cells. Bioconjug Chem 1999, 10:851-860.
26. Vinogradov SV, Bronich TK, Kabanov AV: Self-assembly of polyamine-poly(ethylene glycol) copolymers with phosphorothioate oligonucleotides. Bioconjug Chem 1998, 9:805-812.
27. Brus C, Petersen H, Aigner A, Czubayko F, Kissel T: Physicochemical and biological characterization of polyethyleniminegraft-poly(ethylene glycol) block copolymers as a delivery system for oligonucleotides and ribozymes. Bioconjug Chem2004, 15:677-684.
28. Kunath K, von HA, Petersen H, Fischer D, Voigt K, Kissel T, Bickel U: The structure of PEG-modified poly(ethylene imines) influences biodistribution and pharmacokinetics of their complexes with NF-kappaB decoy in mice. Pharm Res 2002, 19:810-817.
29. Schiffelers RM, Ansari A, Xu J, Zhou Q, Tang Q, Storm G, Molema G, Lu PY, Scaria PV, Woodle MC: Cancer siRNA therapy by tumor selective delivery with ligand-targeted sterically stabilized nanoparticle. Nucleic Acids Res 2004, 32:e149.
30. Fischer D, Osburg B, Petersen H, Kissel T, Bickel U: Effect of poly(ethylene imine) molecular weight and pegylation on organ distribution and pharmacokinetics of polyplexes with oligodeoxynucleotides in mice. Drug Metab Dispos 2004, 32:983-992.
31. Vinogradov SV, Batrakova EV, Kabanov AV: Nanogels for oligonucleotide delivery to the brain. Bioconjug Chem 2004, 15:50-60.
32. Jeong JH, Kim SW, Park TG: A new antisense oligonucleotide delivery system based on self-assembled ODN-PEG hybrid conjugate micelles. J Control Release 2003, 93:183-191.

33. Boussif O, Lezoualc'h F, Zanta MA, Mergny MD, Scherman D, Demeneix B, Behr JP: A versatile vector for gene and oligonucleotide transfer into cells in culture and in vivo: polyethylenimine. Proc Natl Acad Sci U S A 1995, 92:7297-7301.

34. Godbey WT, Wu KK, Mikos AG: Tracking the intracellular path of poly(ethylenimine)/DNA complexes for gene delivery. Proc Natl Acad Sci U S A 1999, 96:5177-5181.

35. Thomas M, Klibanov AM: Enhancing polyethylenimine's delivery of plasmid DNA into mammalian cells. Proc Natl Acad Sci U S A 2002, 99:14640-14645.

36. Sonawane ND, Szoka FC Jr., Verkman AS: Chloride accumulation and swelling in endosomes enhances DNA transfer by polyamine-DNA polyplexes. J Biol Chem 2003, 278:44826-44831.

37. Akinc A, Thomas M, Klibanov AM, Langer R: Exploring polyethylenimine-mediated DNA transfection and the proton sponge hypothesis. J Gene Med 2005, 7:657-663.

38. Sirsi SR, Williams J, Lutz GJ: Poly(ethylene imine)-Polyethylene Glycol Copolymers Facilitate Efficient Delivery of Antisense Oligonucleotides to Nuclei of Mature Muscle Cells of mdx Mice. Hum Gene Ther 2005, 16 (11):1307-1317.

39. Sung SJ, Min SH, Cho KY, Lee S, Min YJ, Yeom YI, Park JK: Effect of polyethylene glycol on gene delivery of polyethylenimine. Biol Pharm Bull 2003, 26:492-500.

40. Glodde M, Sirsi SR, Lutz GJ: Physiochemical Properties of Low and High Molecular Weight PEG-Grafted Poly(ethylene imine) Copolymers and their Complexes with Oligonucleotides. Biomacromolecules 2006, 7(1):347-356.

41. Mann CJ, Honeyman K, McClorey G, Fletcher S, Wilton SD: Improved antisense oligonucleotide induced exon skipping in the mdx mouse model of muscular dystrophy. J Gene Med 2002, 4:644-654.

42. Wells KE, Fletcher S, Mann CJ, Wilton SD, Wells DJ: Enhanced in vivo delivery of antisense oligonucleotides to restore dystrophin expression in adult mdx mouse muscle. FEBS Lett 2003, 552:145-149.

43. Thomas M, Klibanov AM: Conjugation to gold nanoparticles enhances polyethylenimine's transfer of plasmid DNA into mammalian cells. Proc Natl Acad Sci U S A 2003, 100:9138-9143.

44. Shukla R, Bansal V, Chaudhary M, Basu A, Bhonde RR, Sastry M: Biocompatibility of gold nanoparticles and their endocytotic fate inside the cellular compartment: a microscopic overview. Langmuir 2005, 21:10644-10654.

45. Wells DJ, Wells KE, Asante EA, Turner G, Sunada Y, Campbell KP, Walsh FS, Dickson G: Expression of human full-length and minidystrophin in transgenic mdx mice: implications for gene therapy of Duchenne muscular dystrophy. Hum Mol Genet 1995, 4:1245-1250.

46. Phelps SF, Hauser MA, Cole NM, Rafael JA, Hinkle RT, Faulkner JA, Chamberlain JS: Expression of full-length and truncated dystrophin mini-

genes in transgenic mdx mice. Hum Mol Genet 1995, 4:1251-1258.

47. Wells KE, Torelli S, Lu Q, Brown SC, Partridge T, Muntoni F, Wells DJ: Relocalization of neuronal nitric oxide synthase (nNOS) as a marker for complete restoration of the dystrophin associated protein complex in skeletal muscle. Neuromuscul Disord 2003, 13:21-31.

48. Aartsma-Rus A, Bremmer-Bout M, Janson AA, den Dunnen JT, van Ommen GJ, van Deutekom JC: Targeted exon skipping as a potential gene correction therapy for Duchenne muscular dystrophy. Neuromuscul Disord 2002, 12 Suppl 1:S71-S77.

49. Aartsma-Rus A, Kaman WE, Bremmer-Bout M, Janson AA, den Dunnen JT, van Ommen GJ, van Deutekom JC: Comparative analysis of antisense oligonucleotide analogs for targeted DMD exon 46 skipping in muscle cells. Gene Ther 2004, 11:1391-1398.

50. Foster K, Foster H, Dickson JG: Gene therapy progress and prospects: Duchenne muscular dystrophy. Gene Ther 2006, 13:1677-1685.

51. Adams AM, Harding PL, Iversen PL, Coleman C, Fletcher S, Wilton SD: Antisense oligonucleotide induced exon skipping and the dystrophin gene transcript: cocktails and chemistries. BMC Mol Biol 2007, 8:57.

52. Hainfeld JF, Powell RD: New frontiers in gold labeling. J Histochem Cytochem 2000, 48:471-480.

53. Noh SM, Kim WK, Kim SJ, Kim JM, Baek KH, Oh YK: Enhanced cellular delivery and transfection efficiency of plasmid DNA using positively charged biocompatible colloidal gold nanoparticles. Biochim Biophys Acta 2007, 1770:747-752.

54. Bremmer-Bout M, Aartsma-Rus A, de Meijer EJ, Kaman WE, Janson AA, Vossen RH, van Ommen GJ, den Dunnen JT, van Deutekom JC: Targeted exon skipping in transgenic hDMD mice: A model for direct preclinical screening of human-specific antisense oligonucleotides. Mol Ther 2004, 10:232-240.

55. Lu QL, Liang HD, Partridge T, Blomley MJ: Microbubble ultrasound improves the efficiency of gene transduction in skeletal muscle in vivo with reduced tissue damage. Gene Ther 2003, 10:396-405.

56. Lu QL, Bou-Gharios G, Partridge TA: Non-viral gene delivery in skeletal muscle: a protein factory. Gene Ther 2003, 10:131-142.

57. Moulton HM, Fletcher S, Neuman BW, McClorey G, Stein DA, Abes S, Wilton SD, Buchmeier MJ, Lebleu B, Iversen PL: Cell-penetrating peptide-morpholino conjugates alter pre-mRNA splicing of DMD (Duchenne muscular dystrophy) and inhibit murine coronavirus replication in vivo. Biochem Soc Trans 2007, 35:826-828.

58. Lutz GJ, Sirsi SR, illiams JH: PEG-PEI Copolymers for Oligonucleotide Delivery to Cells and Tissues. In Methods in Molecular Biology Volume 9. 433(1) edition. Edited by: Ledoux JM. Humana Press; 2007:141-158.

Chapter 3

GRAPHENE-BASED POLYMER NANOCOMPOSITES

Jeffrey R. Potts [a], Daniel R. Dreyer [b], Christopher W. Bielawski [b*], and Rodney S. Ruoff [a,**]

a Department of Mechanical Engineering and the Texas Materials Institute, The University of Texas at Austin, 204 E. Dean Keeton St., Austin, TX 78712, USA

b Department of Chemistry and Biochemistry, The University of Texas at Austin, One University Station A5300, Austin, TX 78712, USA

ABSTRACT

Graphene-based materials are single- or few-layer platelets that can be produced in bulk quantities by chemical methods. Herein, we present a survey of the literature on polymer nanocomposites with graphene-based fillers including recent work using graphite nanoplatelet fillers. A variety of routes used to produce graphene-based materials are reviewed, along with methods for dispersing these materials in various polymer matrices. We also review the rheological, electrical, mechanical, thermal, and barrier properties of these composites, and how each of these composite properties is dependent upon the intrinsic properties of graphene-based materials and their state of dispersion in the matrix. An overview of potential applications for these composites and current challenges in the field are provided for perspective and to potentially guide future progress on the development of these promising materials.

INTRODUCTION

Graphene, a monolayer of sp2-hybridized carbon atoms arranged in a two-dimensional lattice, has attracted tremendous attention in recent years owing to its exceptional thermal, mechanical, and electrical properties [1e3]. One of the most promising applications of this material is in polymer nanocomposites, polymer matrix composites which incorporate nanoscale filler materials. Nanocomposites with exfoliated layered silicate fillers have been investigated as early as 1950 [4], but significant academic and industrial interest in nanocomposites came nearly forty years later following a report from researchers at Toyota Motor Corporation that demonstrated large mechanical property enhancement using montmorillonite as filler in a Nylon-6 matrix [5]. Polymer nanocomposites show substantial property enhancements at much lower loadings than polymer composites with conventional micron-scale fillers (such as glass or carbon fibers), which ultimately results in lower component weight and can simplify processing [6]; moreover, the multifunctional property enhancements made possible with nanocomposites may create new applications of polymers [7].

On account of the recent emergence of using graphite oxide (GO) to prepare graphene-based materials for composites and other applications [8], this review will focus primarily on polymer nanocomposites utilizing GO-derived materials as fillers. Emphasis will be directed toward structureeproperty relationships as well as trends in property enhancements of these composites, and comparisons to other nanofillers will be made where appropriate.

Some highlights from the literature on polymer composites with what have been referred to as graphite nanoplatelet (GNP) fillers, typically derived from graphite intercalation compounds (GICs), will also be presented and used to provide additional context. Although a review on GO-derived polymer nanocomposites has recently appeared [9], our review considers work with GNP fillers, and provides a historical perspective with more emphasis on preparative methods and processing.

PROPERTIES AND PRODUCTION OF GRAPHENE-BASED MATERIALS FOR COMPOSITE FILLER

Overview and History of Graphene-based Materials

Graphene has a rich history which spans over forty years of experimental work [10]. 'Pristine' graphene (a single, purely sp2-hybridized carbon layer free of heteroatomic defects) has been produced by several routes [1,11], including growth by chemical vapor deposition (both of discrete monolayers onto a substrate and agglomerated powders), micro-mechanical exfoliation of graphite, and growth on crystalline silicon carbide. While these approaches can yield a largely defect-free material with exceptional physical properties, current techniques of making powdered samples of graphene do not yield large enough quantities for use as composite filler [12].

Scalable approaches to GNPs and graphene-based materials (few-layer platelets or monolayer carbon sheets with heteroatoms and topological defects) primarily utilize GICs or GO as the precursor material, respectively. GO and GICs have been investigated as far back as the 1840s [13,14]. In the 1960s, Boehm and coworkers reported the reduction of dispersions of GO using a variety of chemical reductants [15,16], as well as thermal expansion and reduction [17], producing thin, lamellar carbon containing only small amounts of hydrogen and oxygen. By using transmission electron microscopy (TEM), the carbon material produced by chemical reduction was found to consist of "single carbon layers" [15]. The procedures currently used to produce these graphene-like materials have changed little since this early work [18,19]. Techniques for the exfoliation of GICs have also been developed, although most of these approaches do not yield single-layer sheets but rather platelets with thicknesses typically above approximately 5 nm. Liquid-phase exfoliation of graphite and chemical synthesis of graphene from polycyclic aromatic hydrocarbon precursors may also eventually provide scalable alternative routes for production of graphene [19], as could the further development of gas phase CVD methods [12]. Recently, graphene nanoribbons, produced by "unzipping" of multiwalled carbon nanotubes, have been investigated as composite filler [20].

A variety of uses have been envisioned or demonstrated for GNPs and graphene-based materials, and their use as a composite filler has attracted considerable interest [1]. While polymer nanocomposites incorporating GNP fillers continue to be a significant research focus, recent work has largely focused on use of graphene-based filler materials derived from GO. As will be described in the following sections, GO-derived fillers can exhibit high electrical conductivities (on the order of thousands of S/m) [8], high moduli (reported values ranging from 208 GPa [21] to over 650 GPa [22]), and can be easily functionalized to tailor their compatibility with the host polymer.

The reported values of stiffness and electrical conductivity of GOderived filler materials can be higher than those reported for nanoclays [23], but generally lower than those reported for single-walled carbon nanotubes (SWNTs) [24]. However, the intrinsic mechanical properties and electrical and thermal conductivities of SWNTs may be comparable to those of pristine graphene [24,25]. Moreover, the two-dimensional platelet geometry of graphene and graphenebased materials may offer certain property improvements that SWNTs cannot providewhen dispersed in a polymercomposite, such as improved gas permeation resistance of the composite [26].

Exfoliation of Graphite

Most exfoliated graphite fillers are derived from GICs, which are compounds of graphite with atoms or molecules (such as alkali metals or mineral acids) intercalated between the carbon layers [27]. The intercalation of graphite increases its interlayer spacing, weakening the interlayer interactions and facilitating the exfoliation of the GIC by mechanical or thermal methods [28]. Varying structural arrangements of the intercalant are possible, such as alternating layers of graphene and intercalant (referred to as firststage GICs), as well as multiple (two to five) adjacent graphene layers between intercalant layers (higher-stage GICs) [27]. It is the former arrangement, however, which is preferred for the complete exfoliation of these materials into monolayer platelets [29].

Intercalation of graphite by a mixture of sulfuric and nitric acid produces a higher-stage GIC that can be exfoliated by rapid heating or microwave treatment of the dried down powder, producing a material commonly referred to as expanded graphite (EG) [30]. EG retains a layered structure but has slightly increased interlayer spacing relative to graphite, consisting of thin platelets (30e80 nm) which are loosely stacked [30]. Notably, an acid treatment may also oxidize the platelets, albeit to a far lesser degree than GO [31]. EG itself has been investigated as a composite filler [32e34], although its effectiveness in enhancing properties compared with GOderived fillers is limited by its layered structure and relatively low specific surface area (generally less than 40 m2/g [35]). To produce a higher surface area material, EG can be further exfoliated by various techniques to yield GNPs down to 5 nm thickness [28,30].

The thickness and lateral dimensions of GNPs vary widely depending on the production method used, and several approaches have been reported. Sonication of EG in appropriate media can yield platelets with thicknesses of roughly 10 nm and with lateral dimensions as large as 15 mm [36]. Smaller platelet thicknesses have been reported by re-intercalation of EG or co-intercalation of GICs with organic molecules prior to exfoliation. For instance, mixing potassium with EG yielded a stoichiometric first-stage GIC (KC8), which was reported to be exfoliated into GNPs with thicknesses as low as 2 nm upon reaction with water or alcohols, along with sonication [29]. It has also been reported that sulfuric acidintercalated EG can be co-intercalated with tetrabutylammonium hydroxide (among many other molecules [37]). By sonicating this GIC in N,N-dimethylformamide (DMF) in the presence of a surfactant (a poly(ethylene glycol)-modified phospholipid), monolayer graphene-like sheets were obtained [31].

The isolation of pristine graphene by micro-mechanical exfoliation [38] has helped to motivate research towards a scalable procedure for liquid-phase exfoliation of graphite to afford highquality graphene without the use of intercalants; as a result, several different approaches have now been reported [39e42]. Sonication of graphite flakes in water, for example, was reported to yield a mixture of monolayer and multi-layer graphene whichwas largely defect-free, but this approach required the use of surfactants which may negatively affect the electrical conductivity [41]. Direct exfoliation of graphite into solvents such as propylene carbonate (PC) or N-methylpyrrolidone (NMP) [40], and electrochemical exfoliation of graphite in ionic liquids [42] mayeventually offer viable alternatives for production of solution-based graphene, although graphene made by these approaches has yet to be studied as composite filler.

Production and properties of GO

GO is generally produced by the treatment of graphite using strong mineral acids and oxidizing agents, typically via treatment with KMnO4 and H2SO4, as in the Hummers method or its modified derivatives, or KClO3 (or NaClO3) and HNO3 as in the Staudenmaier or Brodie methods [8]. These reactions achieve similar levels of oxidation (C:O ratios of approximately 2:1) [8] which ultimately disrupts the delocalized electronic structure of graphite and imparts a variety of oxygen-based chemical functionalities to the surface.

While the precise structure of GOremains a matter of debate [8], it is thought that hydroxyl and epoxy groups are present in highest concentration on the basal plane, with carboxylic acid groups around the periphery of the sheets as shown in Fig. 1. GO has an expanded interlayer spacing relative to graphite which depends on humidity (for instance, 0.6 nmwhen subjected to high vacuum [43] to roughly 0.8nmat 45% relative humidity [44]) due to intercalation of water molecules [43]. GO can be exfoliated using a variety of methods (most commonly by thermal shocking [45] or chemical reduction in appropriate media [11,46]), yielding a material reported to be structurally similar to that of pristine graphene on a local scale; these techniques will be discussed in more detail below.

Exfoliation of GO

With a few exceptions, production of well-dispersed polymer nanocomposites with GO-derived fillers hinges largely on the exfoliation of GO prior to incorporation into a polymer matrix. Solvent-based exfoliation and thermal exfoliation techniques have emerged as two preferred routes for this step. In the former route, the hydrophilic nature and increased interlayer spacing of GO (relative to graphite) facilitates direct exfoliation intowater assisted by mechanical exfoliation, such as ultrasonication and/or stirring, at concentrations up to 3 mg/ml, forming colloidal suspensions of 'graphene oxide' (that we define with the acronym 'GeO') [11].

Fig. 1 illustrates the structural difference between layered GO and exfoliated GeO platelets. Zeta potential measurements indicate that these suspensions are electrostatically stabilized by negative charges, possibly from the carboxylate groups that are believed to decorate the periphery of the lamellae [11]. Suspensions produced by sonication of GO are found, by atomic force microscopy (AFM) (when deposited onto various substrates), to consist primarily of single-layer GeO platelets [47,48] (Fig. 2); however, the sonication treatment fragments the platelets, reducing their lateral dimensions by over an order of magnitude down to a few hundred nanometers [8,49]. Mechanical stirring is an alternative route to produce singlelayer GeO platelets of much larger lateral dimensions and aspect ratios when compared with GeO platelets produced by sonication.

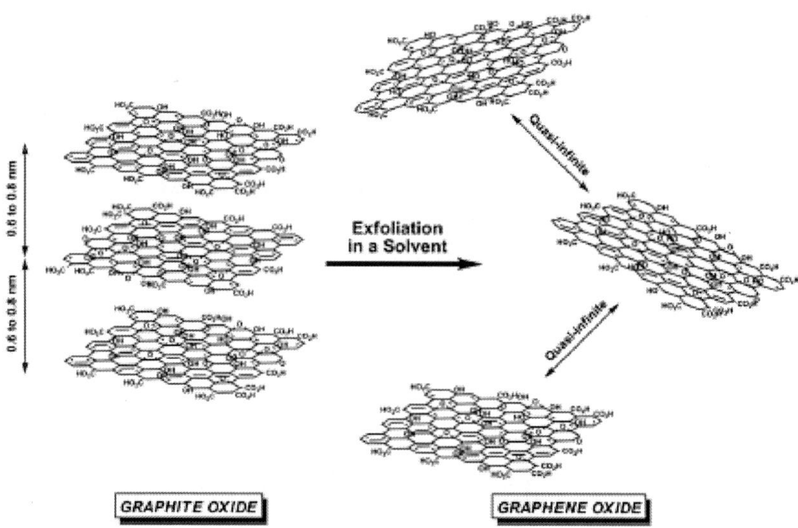

Figure 1. Schematic illustrating the chemical structure of graphite oxide (GO) and the structural difference between layered GO and exfoliated graphene oxide (GeO) platelets.

However, it has been reported that magnetic stirring exfoliates GO very slowly and in low yield [8]. GO can also be exfoliated to GeO platelets of similar aspect ratio to sonicated platelets (at lower concentrations below 0.5 mg/ml) via sonication in polar organic solvents such as DMF, PC, and NMP [50,51].

GO can also be exfoliated and reduced by rapid heating [45], yielding thermally expanded graphite oxide, or TEGO (also referred to commonly in the literature as functionalized graphene sheets, or FGS) [52,53]. In this exfoliation method, the dry powder is typically charged into a quartz tube (or other similar vessel) and subjected to thermal shock (i.e., exposure to a sudden jump in temperature), by heating to temperatures such as 400 C [45] or higher [52] at high rates such as 2000 C/min [52]. The rapid heating is believed to cause various small molecule species (e.g., CO, CO2, water) to evolve and internal pressure to increase, forcing the sheets apart and yielding a dry, high-surface area material with a low bulk density (Fig. 2) [53]. Measurements of the surface area of TEGO by the Brunauer, Emmett, and Teller (BET) method [54] (which measures surface area based on isothermal gas adsorption/desorption) were found to range from 700 to 1500 m2/g [52], compared with a theoretical limit of approximately 2600 m2/g for graphene [55].

Also, GO can be exfoliated (and reduced) by microwave radiation, yielding a related material referred to as microwave-expanded graphite oxide, or MEGO [56]. Importantly, while GeO platelets are likely to maintain the chemical structure of GO, TEGO and MEGO are reduced (reported C:O ratios of 10:1 [52] and 3:1 [56], respectively), and are electrically conductive (reported values of roughly 2000 S/m [52] for TEGO [52] and 270 S/m for MEGO [56]). TEGO was reported to contain residual oxygen (in the form of carbonyl and ether groups) and adopted a crumpled accordion-like morphology, with lateral dimensions of a fewhundred nanometers, similar to GeO platelets exfoliated by sonication [52,57].

Chemical reduction and functionalization of GeO platelets

The physical properties of GeO platelets are considerably different from that of graphene. GeO platelets can be chemically reduced to generate a material that resembles pristine graphene, using reducing agents such as hydrazine monohydrate or sodium borohydride

[11]. Chemically reduced GeO platelets (henceforth, wewill use the acronymRGeO) can exhibit C:O ratios of over 10:1 [8] and retain some of the functionality originally present on the GeO platelets [11,58]. Recently, it was reported that GO or GeO platelets may be used to catalyze the oxidation of a variety of benzylic and aliphatic alcohols [59], as well as various carbonecarbon bond forming reactions [60], while reducingGOaswell asGeOplatelets in the process. The products recovered after reduction by benzyl alcohol were reported to exhibit high C:O ratios (up to 29.9:1) and high electrical conductivities (up to 4600 S/m) [61]. Other reductionmethods using environmentally friendly reductants, such as tryptophan and ascorbic acid, have been reported [62,63].

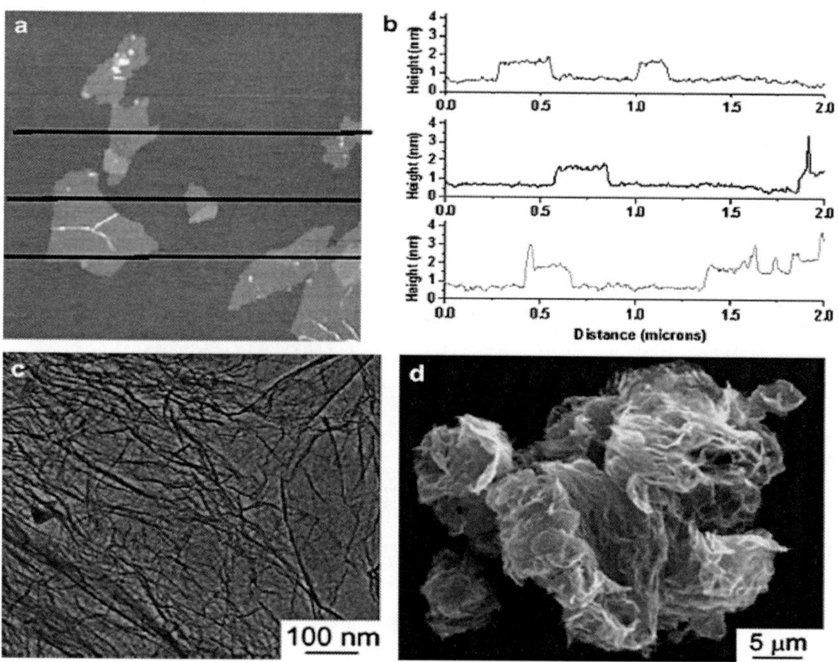

Figure 2. (a, b) Non-contact AFM scans of graphene oxide (GeO) deposited on mica reveal the presence of single layers obtained from exfoliation in water via sonication. The wrinkled structure of thermally expanded GO (TEGO) is illustrated in this transmission electron micrograph (c) and scanning electron micrograph (d) ((a) and (b) were adapted from Ref. [64], (c) and (d) were adapted from Ref. [53]).

This reduction process can cause agglomeration of the platelets [64] (reducing accessible surface area) unless prior steps are taken to stabilize the suspension. Adjusting the pH of the suspension to increase the (negative) zeta potential of the sheets or the adsorption of polymers on the platelet surface are both effective routes to stabilizing aqueous suspensions of RGeO platelets [65,66]. Stable suspensions of RGeO platelets in organic solvents have also been achieved. One approach to these suspensions is progressive dilution of an aqueous suspension of GeO platelets with an organic solventdwith one possible route involving hydrazine in DMF:water (9:1 v/v) and further dilution in DMF to yield a stable suspension of RGeO platelets in 99% DMF [46]. Twophase extraction of RGeO platelets from water into various organic solvents may be facilitated by end-functional polymers dissolved in the organic phase, which adsorb onto the platelets and help disperse the platelets in the solvent [67]. Freeze drying of aqueous RGeO dispersions has been reported to facilitate the redispersion of the RGeO platelets into organic solvents such as DMF [68]. Alternatively, onemay avoid the use of water altogether,

as stable dispersions of RGeO platelets in DMF and NMP have also been reported using dimethylhydrazine as the reductant [69]. For composites processed in solution, chemical reduction of GeO platelets in the polymer solution (provided the polymer is stable to the reaction conditions) may prevent the precipitation of the RGeO platelets from the solvent, as the polymer can maintain the dispersion of RGeO platelets [55].

While the most commonly employed reaction of GeO is its reduction to yield electrically conductive RGeO, a variety of other chemical transformations can be carried out at its oxygen-based functional groups, which are covered elsewhere [8]. Both covalent and non-covalent functionalization of GeO platelets has been reported to generate stable dispersions of chemically modified graphene (CMG) platelets in organic solvents and also to enhance their compatibility with various polymer matrices. Among others, reactions using amines [70e72] and isocyanates [73] have been reported for small molecule functionalization of GeO platelets because of the facility of the reactions, and the ability to react in multipleways (e.g., amidations, nucleophilic epoxide ring-openings, carbamate formation, etc.) However, covalent functionalization of

GeO platelets could adversely affect the electrical conductivity of the platelets as these functionalizations disrupt (or retain the disruption already present in) the sp2-hybridized network required for good electron/hole conduction [74]. Non-covalent functionalization of RGeO platelets via, for example, pep stacking could minimize disruption of the conductive, conjugated structure [75,76].

PREPARATION OF GRAPHENE-BASED POLYMER NANOCOMPOSITES

Overview and historical perspective

The earliest reports on polymer composites with exfoliated graphite fillers emerged from studies on the intercalation chemistry of GICs. In 1958, it was discovered that alkali metal-GICs could initiate the polymerization of ethylene [77], and subsequently, alkali metal-GICs were found to initiate polymerization of other monomers such as styrene, methyl methacrylate, and isoprene [78,79].

Early reports focused on the characterization of the polymer produced from these reactions; it was decades later before the observation was reported that alkali metal-GIC-initiated polymerization could exfoliate the layers of the graphite host [37,80].

Building on his own work on the exfoliation of graphite, Bunnell proposed the production of polymer nanocomposites incorporating "as thin as possible" GNPs (derived from GICs exfoliated either by shear grinding or thermal treatment) as fillers in a 1991 patent [81], where he suggested that with "10 vol% inclusion of graphite flakes in.polyethylene or polypropylene, the stiffness of the finished product will approach that of aluminum." However, it was not until 2000 that a detailed study of the morphology and properties of an exfoliated graphite nanocomposite was published, which reported dispersed platelets of approximately 10 nm thickness, produced by exfoliation of EG due to the in situ polymerization of caprolactam [82]. This and subsequent studies have reported tremendous property improvements versus conventional polymer composites based on micron-scale fillers such as untreated flake graphite or carbon black (CB) [83e87]. For instance, much lower electrical percolation thresholds have been reported with GNP fillers versus CB: 8 wt% for CB/PMMA [88] and 9 wt% for CB/Nylon [89], compared with 1 wt% for GNP/PMMA [86] and 1.8 wt% for GNP/Nylon [82].

As with GICs, GO can be intercalated by various monomers, and subsequent polymerization has been reported to delaminate the layers [90]; however, it has also been reported that polymers can be directly intercalated into GO [91e93]. Hydrazine and electrochemical reduction of layered GO/polyelectrolyte films was used to generate electrically conductive polymer composites in 1996 [94], but it was nearly ten years later that an electrically conductive poly (styrene) composite was prepared by using well-dispersed, monolayer CMG fillers [55], stimulating an intense research effort on polymer composites with dispersed CMG platelets (and other GO-derived materials such as TEGO) as fillers.

In recent years, a variety of processing routes have been reported for dispersing both GNP and GO-derived fillers into polymer matrices. Many of these procedures are similar to those used for other nanocomposite systems [95], although some of these techniques have been applied uniquely to graphene-based composites.

Among other factors, the nature of the bonding interaction at the interface between the filler and matrix has significant implications for the final composite properties, and most dispersion methods produce composites that are non-covalent assemblies where the polymer matrix and the filler interact through relatively weak dispersive forces. However, there is a growing research focus on introducing covalent linkages between graphene-based filler and the supporting polymer to promote stronger interfacial bonding, as will be illustrated in the following sections.

Non-covalent dispersion methods: solution and melt mixing

Solution-based methods generally involve the mixing of colloidal suspensions of GeO platelets or other graphene-based materials with the desired polymer, either itself already in solutionor by dissolving in the polymer in the suspension of GeO platelets, by simple stirring or shear mixing. The resulting suspension can then be precipitated using a non-solvent for the polymer, causing the polymer chains to encapsulate the filler upon precipitation. The precipitated composite can then be extracted, dried, and further processed for testing and application. Alternatively, the suspension can be directly cast into a mold and the solvent removed. However, this latter technique can potentially lead to aggregation of the filler in the composite, which may be detrimental to composite properties [95].

Solution mixing has been widely reported in the literature, as CMG platelets can often be processed in either water or organic solvents. This approach has been used for incorporating GOderived fillers into a variety of polymers, including: PS [55,96], polycarbonate [97], polyacrylamide [98], polyimides [99], and poly (methyl methacrylate) (PMMA) [7,100]. The facile production of aqueous GeO platelet suspensions via sonication makes this technique particularly appealing for water-soluble polymers such as poly(vinyl alcohol) (PVA) [100e105] and poly(allylamine), composites of which can be produced via simple filtration [104,106]. In addition, vacuum filtration of GeO/PVA and GeO/PMMA solutions has been used to make composite films across a broad range of loadings [107], which have a layered morphology similar to that of 'graphene oxide paper' [44].

While some restacking of the platelets may be possible, for solution mixing methods the dispersion of platelets in the composite is largely governed by the level of exfoliation of the platelets achieved prior to, or during, mixing. Thus, solution mixing offers a potentially simple route to dispersing single-layer CMG platelets into a polymer matrix. As previously mentioned, small molecule functionalization and grafting-to/from methods have been reported to achieve stable CMG platelet suspensions of highly exfoliated platelets prior to mixing with the polymer host. Lyophilization methods [68], phase transfer techniques [67,108], and surfactants [109] have all been employed to facilitate solution mixing of graphene-based composites. However, the use of surfactants may affect composite properties; for instance, surfactants have been reported to increase the matrixefiller interfacial thermal resistance in SWNT/polymer composites, attenuating the thermal conductivity enhancement relative to SWNTs that were processed without surfactants [110].

In melt mixing, a polymer melt and filler (in a dried powder form) are mixed under high shear conditions. Relative to solution mixing, melt mixing is often considered more economical (because no solvent is used) and ismore compatible withmany current industrial practices [111]; however, studies suggest that, to date, such methods do not provide the same level of dispersion of the filler as solvent mixing or in situ polymerization methods [26]. Notably, no means of dispersing single- or few-layer GO-derived fillers via melt mixing without prior exfoliation have been reported akin to layered silicate fillers (although, with a few exceptions, direct exfoliation of layered silicates in melt mixing requires prior treatment with a surfactant to increase miscibility with the polymer host [112]). Several studies report melt mixing using TEGO [113] and GNPs [114e117] as filler, where these materials could be fed directly into an extruder and dispersed into a polymer matrix without the use of any solvents or surfactants. Notably, the very low bulk density (approximately 0.004 g/cm3 based on a volumetric expansion of 500 [52]) of TEGO makes handling of the dry powders difficult and poses a processing challenge (such as for feeding into processing equipment such as a melt extruder), and in one study a solution mixing process was used to disperse the TEGO in the polymer prior to compounding in order to circumvent this issue [118]. In a different approach to 'premix' the polymer and filler prior to mixing, GNPs were sonicated in a non-solvent, such that polymer particles were uniformly coated with GNPs prior to melt mixing, which was reported to lower the electrical percolation threshold of a GNP/polypropylene composite [119]. Notably, for composites incorporating GeO platelets as filler, melt processing and molding operations may cause substantial reduction of the platelets due to their thermal instability [120].

Non-covalent in situ Polymerization

In situ polymerization methods for production of polymer composites generally involve mixing of filler in neat monomer (or multiple monomers), or a solution of monomer, followed by polymerization in the presence of the dispersed filler. These efforts are often followed with precipitation/extraction or solution casting to generate samples for testing. Many reports using in situ polymerization methods have produced composites with covalent linkages between the matrix and filler, and many examples will be given in the following sections. However, in situ polymerization has also been used to produce non-covalent composites of a variety of polymers, such as poly(ethylene) [121], PMMA [122], and poly (pyrrole) [123,124].

Unlike what has been reported for solution mixing methods, a high level of dispersion of graphene-based filler has been achieved via in situ polymerization without a prior exfoliation step. In some reports, monomer is intercalated between the layers of graphite or GO, followed by polymerization to separate the layers.

This technique, sometimes referred to as intercalation polymerization, has been widely investigated for nanoclay/polymer composites [112], and has been also applied to GNP and GO-derived polymer composites. For instance, in situ polymerization methods have been used to exfoliate GICs and EG to generate dispersions of GNPs in the matrix. Graphite can be intercalated by an alkali metal and a monomer (e.g., isoprene or styrene), followed by polymerization initiated by the negatively charged graphene sheets [37].

However, it is not known if the polymerization takes place on the surface of the GIC or between layers [37]. In any case, in situ polymerization in the presence of GICs has been reported to exfoliate the GIC into thin platelets [80], and this approach has also been reported to exfoliate EG [82,84,85], although exfoliation to afford isolated monolayers has yet to be achieved with this approach. In a recent study, metallocene-mediated polymerization of poly (ethylene) was conducted in the presence of dispersed GNPs, in an attempt to grow PE chains between the graphitic layers. Although the polymerization may have further exfoliated the GNPs, monolayer graphene platelets were not observed; TEM observations showed platelets down to 3.6 nm thickness (consistent with stacks of approximately 10 layers) with relatively low aspect ratios of about 30 dispersed in the PE matrix [121].

The larger interlayer spacing of GO (between about 0.6 and 0.8 nm depending on relative humidity) compared to graphite (0.34 nm) facilitates intercalation by both monomers and polymers [28]. Additionally, the polar functional groups of GO promote direct intercalation of hydrophilic molecules, with the interlayer spacing increasing with uptake of monomer or polymer (e.g., increasing up to 2.2 nm for the intercalation of PVA into GO) [93]. The interlayer spacing of vinyl acetate-intercalated GO was reported to decrease after polymerization [125], although the interlayer spacing still remains significantly higher than unmodified GO. In situ polymerization has been demonstrated for several GO composite systems, including poly(vinyl acetate) [125], and poly(aniline) (PANI) [90].

X-ray diffraction studies on these systems suggested an intercalated morphology where the individual graphene oxide sheets remain loosely stacked in the matrix with polymer intercalated between lamellae. However, the use of a macroinitiator to intercalate GO prior to in situ polymerization of methyl methacrylate was recently reported to improve the filler dispersion in a GO/PMMA composite.

X-ray diffraction revealed an increased interlayer spacing of GO (from 0.64 nm to 0.8 nm) suggesting an intercalated morphology for GO/PMMA composites produced by conventional free radical polymerization. However, composites polymerized with the macroinitiator showed improved reinforcement and no diffraction peak from GO, suggesting a more exfoliated morphology [122].

Graphene-based Composites with Covalent Bonds Between Matrix and Filler

Given the relative sparseness of usable functionality on pure carbon materials, forming covalent linkages between the polymer matrix and such surfaces (when used as a composite filler) may be quite challenging. However, GeO platelets contain a surface rich in reactive functional groups, and a number of approaches for introducing covalent bonds between GeO platelets and polymers have been demonstrated. For instance, both grafting-from and graftingto approaches have been used for the attachment of a broad range of polymers.

In one recent example of a grafting-from approach [126], atom transfer radical polymerization (ATRP) initiators were covalently attached via esterification with the alcohols present across the GeO platelet surface. Upon adding an ATRP-compatible monomer (e.g., styrene, butylacrylate, or methyl methacrylate) and a source of CuI, polymer brushes were grown in a controlled fashion from the surface (Fig. 3). The ester linkage between the polymer and the CMG platelet surface was then saponified using aqueous NaOH, allowing for characterization of the polymers independent of the carbon material. Similar studies using such ATRP-based methods have reported an increased scope of monomer reactivity [96,127], as well as the incorporation of these polymer-grafted CMG platelets into a polymer matrix via solution mixing, reportedly leading to improvements in mechanical and thermal properties versus the neat matrix polymer [96,128,129]. Recent efforts have focused on correlation of thermal properties of these composites as a function of grafting density and polymer molecular weight [130].

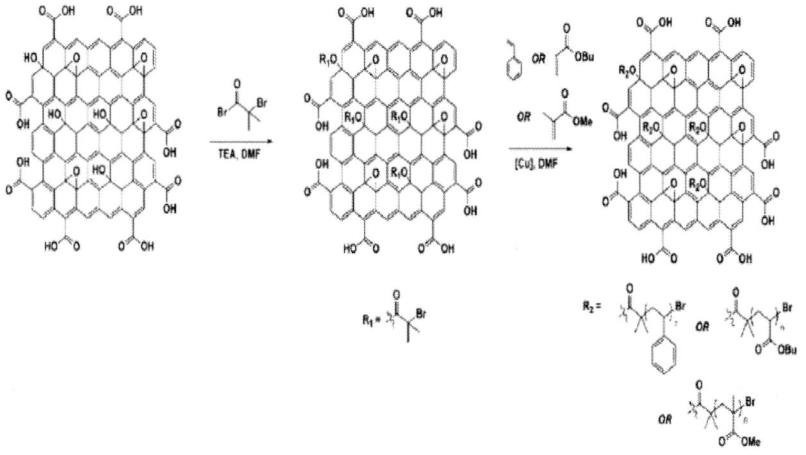

Figure 3. Synthesis of surface-attached poly(styrene), poly(methyl methacrylate), or poly(butylacrylate) via ATRP following functionalization of GeO platelets with an ATRP initiator (a-bromoisobutyryl bromide) (adapted from Ref. [126]).

These grafting-based approaches have also been used in conjunction with heterogeneous blending of polymer-functionalized GeO in matrices composed of conducting polymers [131], including poly(3-hexylthiophene) (P3HT) and a triphenylaminebased poly(azomethine) (see Fig. 4) [132,133]. Both polymers have been widely studied as conducting materials for use in photovoltaics and data storage devices, among other applications, and the incorporation of GeO platelets or other GO-derived materials into these devices may enhance the optoelectronic properties of the device, as well as enhance the device's mechanical and/or thermal properties.

These two examples are good case studies for divergent pathways to polymer-functionalized GeO platelet composites: in the P3HT system, poly(t-butylacrylate) was grown via ATRP from the surface of GeO platelets (grafting-from, where a polymer is grown from a heterogeneous surface), and blended with preformed P3HT for the formation of organic electronic memory devices. Conversely, the poly(azomethine) was attached using the amino functional groups that were pendant to the end groups of the polymer (Fig. 4; grafting-to), possibly through amide formation with carboxylic acid groups present on the edges of GeO platelets.

However, the presence of other reactive sites (e.g., epoxides) may make precise determination of the reactive site difficult [134]. Other reports of grafting-to approaches include reports of grafting of azide-terminated poly(styrene) (PS) chains to the surface of alkyne-functionalized GeO platelets via a CuI-catalyzed 1,3-dipolar cycloaddition in an example of click chemistry [135], and grafting of PVA to GeO platelets via carbodiimide-activated esterification [136]. The choice between using grafting-from or grafting-to methodologies will likely depend on the polymer being formed and which of the two approaches is more practical for the application under consideration. However, a grafting-to approach may lower the grafting density of chains to the platelet surface [137], which may in turn affect the dispersion of these polymer-grafted platelets if dispersed into a polymer matrix [138].

For certain polymers, covalent bonding between the matrix and GeO platelets may form during polymerization (on reaction with the functional groups of GeO) without the need for prior functionalization or controlled grafting methods. For an epoxy matrix composite, curing with an amine hardener may have resulted in the incorporation of GeO platelets directly into the crosslinked network [139], while for polyurethanes, TEGO was reported to function as a chain extender by reacting with the isocyanate groups of the monomer or prepolymer [26,140]. Ring-opening polymerization of caprolactam was reported to graft polyamide brushes to GeO platelets via condensation reactions between the aminecontaining monomer and the carboxylic acid groups of the GeO platelets, though increased loadings of filler were found to lower the polymer molecular weight due to stoichiometric imbalance during polymerization [141].

GeO platelets have also been utilized as a ZieglereNatta catalyst support for the heterogeneous in situ polymerization of propylene. Functionalization of GeO platelets with a Grignard reagent (BuMgCl) served to immobilize TiCl4 on the GeO platelet surface (as evidenced by IR spectroscopy, X-ray photoelectron (XPS) spectroscopy, and energy-dispersive X-ray (EDX) spectroscopy).

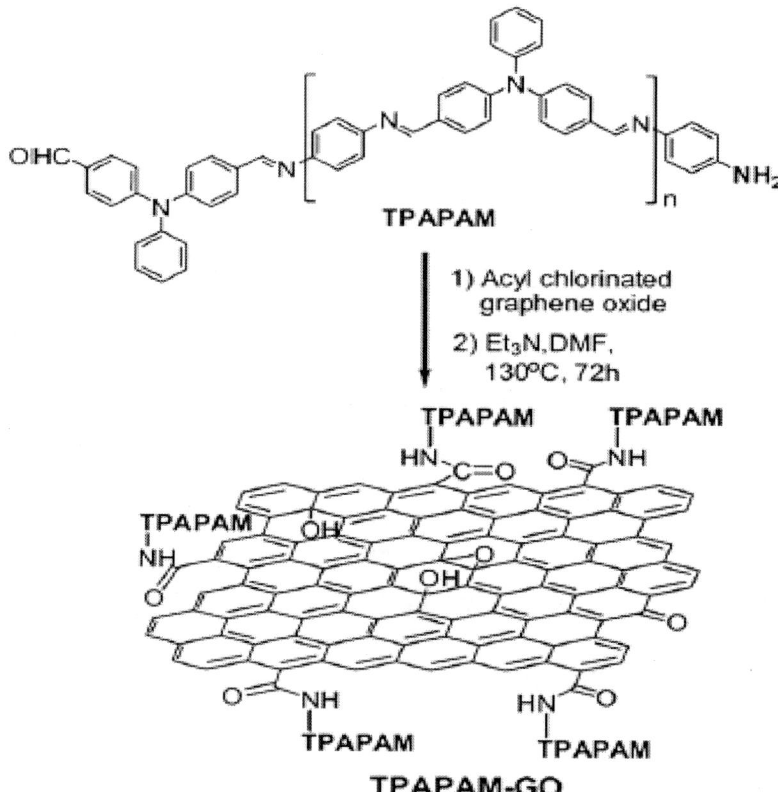

Figure 4. Attachment of a triphenylamine-based poly(azomethine) (TPAPAM) to GO via amidation chemistry, embodying a grafting-to approach (adapted from Ref. [133]).

Subsequent initiation with AlEt3 was reported to afford high molecular weight, isotactic poly(propylene) (Fig. 5). The resulting composites showed a homogeneous dispersion of few-layer CMG platelets, and exhibited moderate electrical conductivity (0.3 S/m at

4.9 wt%), with the authors noting that the GeO platelets were not intentionally reduced [142].

Other Methods for Composite Preparation

In addition to those described above, several other methods have been reported for producing graphene-based composites. Although most of these procedures have been demonstrated on just one composite system, many could potentially find use as general approaches to composite fabrication. One such approach is the noncovalent grafting of well-defined polymers to RGeO platelets via pep interactions. For instance, the attachment of pyrene-terminated poly(N-isopropylacrylamide) to RGeOwas recently reported; the compositewas stated to retain the thermoresponsive properties of the neat polymer [143]. This technique has since been extended to various other polymers [144], suggesting that non-covalent grafting of polymers to the surface of CMG platelets may provide a versatile approach to producing graphene-based composites.

Moreover, such non-covalent composites may better preserve the conjugated structure of graphene-based materials as compared with covalent functionalization or grafting approaches, which may benefit composite properties such as electrical conductivity.

Variants of typical in situ polymerization and solution mixing methods may provide useful methods of dispersing graphenebased fillers in a polymer matrix. For instance, emulsion polymerizations can be carried out in aqueous suspensions of GeO platelets [145,146], suggesting a general approach for dispersion of CMG platelets with latex-based polymers [147]. As previously mentioned, lyophilization methods [68] or phase transfer techniques [67,108] may offer general approaches to dispersing RGeO platelets as filler in a polymer matrix. In one report, reduction of an aqueous suspension of GeO platelets with hydrazine resulted in the extraction of the hydrophobic RGeO platelets into an organic layer (containing the dissolved polymer) and formation of a homogeneously dispersed nanocomposite [108].

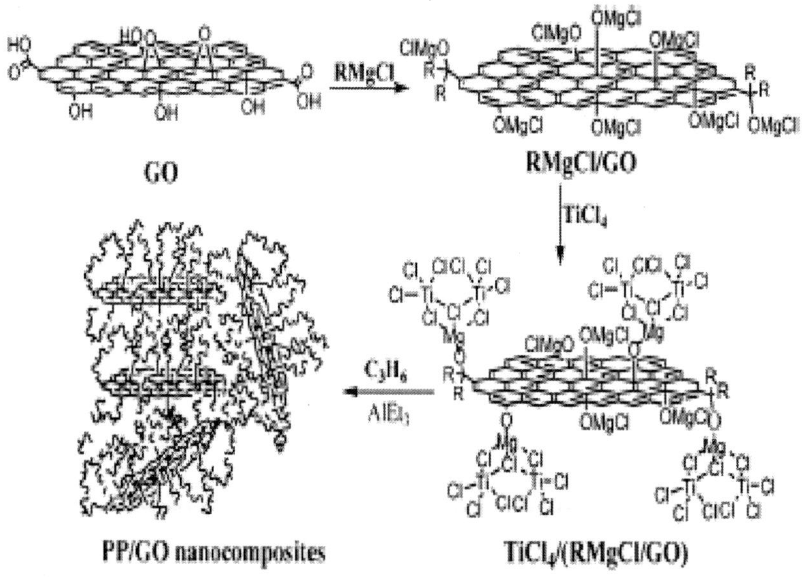

Figure 5. ZieglereNatta polymerization of propylene from the surface of GeO, which acts as a support for the growing polymer chains (adapted from Ref. [142]).

A variety of other methods for composite production have been reported. Attempts to exfoliate graphite directly via conventional melt mixing techniques have not been successful to date [148].

However, solid state shear pulverization, which uses a twin screw extruder to blend solid materials using shear, was reported to exfoliate and disperse unmodified graphite directly into polypropylene, yielding nanocomposites with platelets having thicknesses of approximately 10 nm or less [149]. Other production methods, such as layer-by-layer assembly of polymer composite films [150] and backfilling of GeO platelet aerogel structures (produced by freeze drying aqueous GeO suspensions) with polymer may provide means to produce nanocomposites with defined morphologies [151]. In one study, directional freeze drying of an aqueous GeO platelets/ PVA platelets mixture was reported to yield nanocomposites with a three-dimensional macroporous structure and a surface area of approximately 37 m2/g [152].

MORPHOLOGY AND CRYSTALLIZATION BEHAVIOR

As property enhancements correlate strongly with nanocomposite microstructure, effective characterization of morphology is important to establishing structureeproperty relationships for these materials. For instance, TEM of microtomed thin sections of the composite can provide direct observation of dispersed multilayer GNPs and graphene-based platelets; such thicker platelets typically show adequate contrast against the polymer matrix to be imaged without staining, whereas single-layer platelets may be difficult to directly observe by TEM [26]. Compared with TEM, wide-angle X-ray scattering (WAXS) can more rapidly provide insight into the state of dispersion over a larger volume of composite; however, since the scattering intensity varies with the concentration of the scattering feature, some morphological information may be missed [148].

Both graphite and GO, as the precursors to many graphenebasedmaterials, have a layered structure as do certain silicates (e.g., montmorillonite) which have been widely investigated as composite fillers [111]. Indeed, when dispersed into a polymer matrix, both nanoclays and graphene-based platelets exhibit similar states of dispersion depending upon factors such as the processing technique and the affinity between the phases. Moreover, nanoclay fillers often exhibit comparable aspect ratios to graphene-based fillers (up to 1000) [112], although fillers such as TEGO often appear more crumpled on a local scale relative to nanoclays [7]. Earlier studies on nanoclay-based composites have suggested the existence of three general states of platelet dispersion on short length scales: stacked, intercalated, or exfoliated, as shown in Fig. 6. As similar morphologies have been observed in the literature on both GO-derived and GNP/polymer nanocomposites, we thus suggest extension of this terminology to these systems.

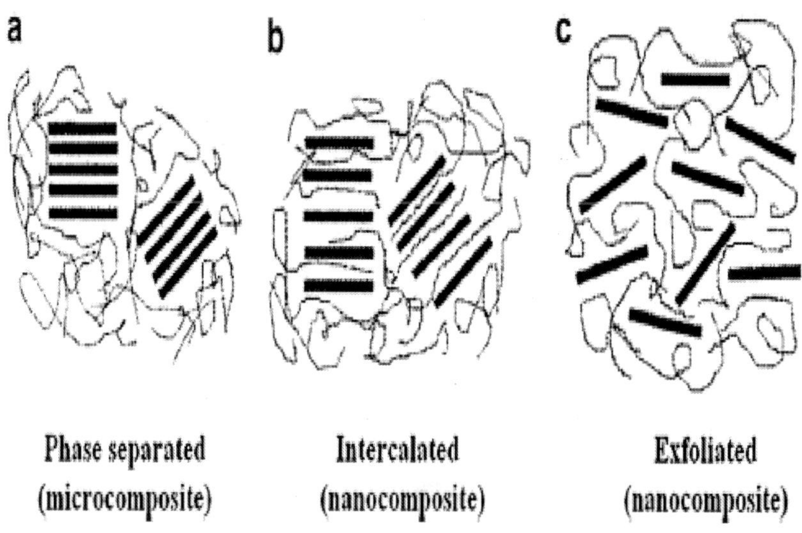

Figure 6. Schematic showing three morphological states, as originally suggested for layered silicate fillers, that are also possible with graphene-based nanocomposites: (a) phase separated, (b) intercalated, (c) exfoliated (adapted from Ref. [23]).

TEM and WAXS studies are perhaps the two most common means by which the state of dispersion can be assessed. Immiscibility of the phases and/or insufficient exfoliation of the graphite or GO-derived filler prior to mixing with polymer can result in large agglomerates consisting of stacked platelets when observed by TEM, which may also be suggested by the presence of a diffraction peak corresponding to the interlayer spacing of GO or graphite [26,148,153]. Intercalated platelets retain a stacked structure but with increased interlayer spacing (on the order of a few nanometers), as evidenced by a shifted diffraction peak from that of unmodified graphite or GO [154]. As will be discussed in the following sections, high aspect ratio platelets are generally found to be beneficial to the mechanical, electrical, and thermal properties of a composite material. An exfoliated morphology of GO or GICs is thus usually desired as it provides higher aspect ratio platelets relative to stacked or intercalated platelets [155]. This state of dispersion may be suggested by a scattering profile corresponding to that of the neat matrix polymer; however, multi-layer intercalated platelets could actually be dispersed (as observed by TEM) despite the absence of a diffraction peak. Conversely, quantitative evaluation of platelet exfoliation and geometry via TEM poses its own set of challenges (such as sampling sufficient number of filler particles at such high magnification and the possible influence of TEM sample preparation on apparent level of dispersion).

WhileWAXSorTEMcan be used to assess dispersion of individual platelets, neither can detect larger-scale morphological features [111,156]. Small-angle X-ray scattering (SAXS) and ultra-small-angle X-ray scattering (USAXS) measurements have been used on a variety of nanocomposite systems to detect the presence of fractal-like aggregates of filler at length scales beyond that of individual particles, although only limited information of this nature exists on GOderived polymercomposites [148], perhaps due in part to the limited accessibility of such techniques [111].However, TEGO/polycarbonate nanocomposites were recently examined by small-angle neutron scattering. These measurements were used to quantify dispersion of the platelets (based on an idealized platelet model developed for montmorillonite), which suggested a decreasing effective aspect ratio with increased loading of TEGO and increased aggregation of filler at higher loadings [157]. Finally, cross-sectional analysis with scanning electron microscopy (SEM) has been used to evaluate dispersion of graphene-based filler [55] as well as to examine the surface for filler pull-out, possibly giving insight into the strength of interfacial adhesion [7,158]. However, care must be exercised when identifying the dispersed filler; moreover, SEM generally cannot resolve the degree of exfoliation of the platelets and is therefore best utilized as a complementary technique.

Exfoliated graphene-based materials are often compliant, and when dispersed in a polymer matrix are typically not observed as rigid disks, but rather as bent or crumpled platelets. Moreover, graphene has been shown to 'scroll up' irreversibly when its polymer host is heated above its glass transition temperature (Tg) [159]. Compatibility between the polymer matrix and the CMG platelets also can reportedly affect the platelets' conformation [160]. If the platelets' affinity for the matrix is high, then the particles may adopt a more extended conformation. However, the platelets may gradually adopt a more crumpled conformation as the affinity between the components decreases [160]. The technique used to process the composites can affect the microstructure, as shown in Fig. 7: randomly oriented, exfoliated platelets may be favored when composites are processed by solution mixing or in situ polymerization, compared with a more oriented and intercalated/stacked structure for composites produced by melt mixing, possibly due to restacking of the platelets [26]. The processing technique can also induce orientation of the dispersed platelets, which can be beneficial for reinforcement [161] but may raise the percolation threshold [153]. For composites processed by injection molding, platelets may be more randomly oriented near the interior of the specimen, with platelets aligned parallel to the surface [153]. By comparison, sufficiently thin, compression molded specimens [153] or solution-cast films [99] may have aligned platelets along the entire cross section. The filler type may also affect the orientation of the dispersed platelets. As shown in Fig. 8, solutioncast TEGO/Nafion composites exhibited a randomly oriented dispersion of TEGO platelets, whereas solution mixing of Nafion with GeO platelets, followed by hydrazine reduction, produced a highly oriented, uniform dispersion of RGeO platelets (it was stated that the reduction did not affect the orientation) [162]. The consequences of platelet conformations and orientations on composite properties will be discussed in more detail below.

For semicrystalline polymers, incorporation of a nanofiller can lead to an altered degree of crystallinity, crystallite size, spherulite structure, and may even induce crystallization of otherwise amorphous polymers [163,164]. Depending on the identity of the polymer, incorporation of graphene-based filler has been reported to cause increases [101,103,165,166], decreases [167], or no change [162] in the degree of crystallinity of a semicrystalline polymer matrix; changes in the polymer melting temperature have also been reported [168]. The presence of graphene-based nanofillersmay also affect the rate of crystallization, by serving as a heterogeneous nucleation site for crystal growth [163]. Additionally, GNPs have been reported to accelerate the crystallite growth kinetics of poly (L-lactide), though the effect was found to be less pronounced than with carbon nanotubes (CNTs) [164].

Aside from crystallization, incorporation of graphene-based filler can impart other changes in the morphology of the polymer matrix or composite structure. The morphology of a self-assembling triblock copolymer, poly(styrene-block-isoprene-blockstyrene), was stated to be affected by the presence of TEGO filler:

AFM and electrostatic force microscopy (EFM) studies on approximately 300 nm-thick films prepared by spin coating revealed loss of long range order in the domains, with the TEGO preferentially dispersed in the PS blocks where they adopted a folded conformation [169]. Highly aligned GeO platelets dispersed in Nafion may have directed the orientation of the ionic domains of Nafion parallel to the surfaces of solution-cast films of the composites [162]. For electrospun graphene-based composites, a poor dispersion of CMG platelets has been reported to induce formation of bead-like structures in the fibers [170].

RHEOLOGICAL AND VISCOELASTIC PROPERTIES

Study of nanocomposite rheology is important for the understanding of processing operations but it may also be used to examine nanocomposite microstructure [171e173]. In linear viscoelastic rheology measurements, the low-frequency moduli may provide information on the platelet dispersion; for instance, the presence of a low-frequency storage modulus (G0) plateau is indicative of rheological percolation due to formation of a 'solidlike' elastic network of filler [174]; an example is illustrated in Fig. 9 for a TEGO/ polycarbonate composite. The onset of a frequencyindependent G0 may also coincide with other phenomena, such as the loading at which a large decrease in the linear viscoelastic strain limit is observed [153]. The percolation threshold determined from such measurements can be used to roughly quantify dispersion in terms of an equivalent aspect ratio of idealized platelets [148,153].

Generally, G0 has been found to increase across all frequencies with dispersion of rigid nanoplatelets, consistent with reinforcement. In addition to melt rheology, changes in the dynamic moduli have been studied in several composite systems with GNP and GOderived fillers using dynamic mechanical analysis (DMA) temperature scans [175e178].

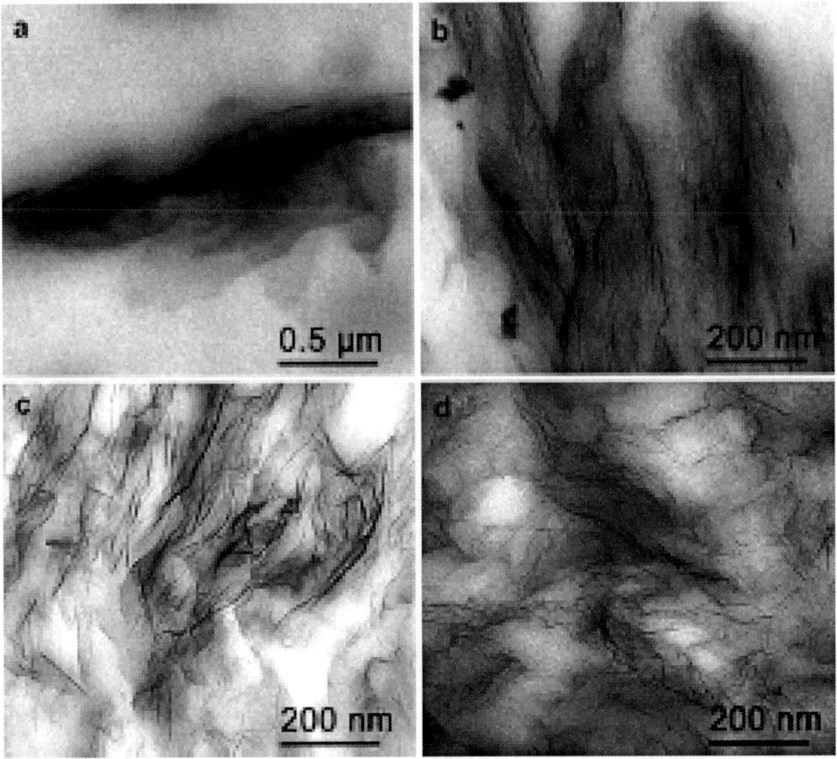

Figure 7. TEM images illustrating the morphological differences in composites with a thermoplastic poly(urethane) matrix filled with (a) unexfoliated graphite in a stacked morphology, and (b) TEGO, processed by melt mixing. Images (c) and (d) show TEGO/polyurethane composites produced by solution blending and in situ polymerization, respectively, illustrating a more exfoliated state of dispersion (adapted from Ref. [26]).

As previously mentioned, orientation of CMG platelets has been stated to affect the onset of rheological percolation, as randomly oriented, well-dispersed platelets would be expected to percolate at lower concentrations than aligned, well-dispersed platelets. One route to promoting the randomization of filler orientation is thermal annealing above the Tg of the polymer. As Fig. 9 shows, the rheological percolation threshold of a TEGO/polycarbonate composite was lowered from 1.5 vol% to 0.5 vol% by annealing for several hours [153]. Moreover, orientation of the platelets (induced by high strain) lowered the melt elasticity, while subsequent annealing steps were reported to restore the solid-like behavior of the composite melt. Hence, annealing of the composites following molding operations may provide a route to improve properties that benefit from randomly oriented (rather than aligned) platelets, such as the percolation threshold for electrical conductivity.

Lower composite solution viscosities have been reported with GNP fillers compared to CNTs [179], which may be advantageous for solution-based processing techniques, such as commercial molding processes for epoxy composite thermal interface materials. It has been suggested that at sufficiently high loadings, entanglement of CNTs in the matrix could result in undesirably large viscosity increases, whereas platelets can more easily slide past one another, thus moderating the viscosity increase [28]. Nonetheless, the solution viscosities of CMG/epoxy composites have been found to increase substantially with loading of filler, which could inhibit the formation of the crosslinked epoxy network [180]. It has been reported that functionalization to enhance compatibility of the filler with the polymer matrix may help to moderate the composite solution viscosity with increased loading [175]. Notably, GO composite solutions may exhibit electro-rheological properties, a characteristic of insulating colloidal particles in insulating media where increases in solution viscosity due to morphological changes can be observed upon application of an electric field [181].

CHANGES IN THE GLASS TRANSITION TEMPERATURE

Low loadings of CMG fillers have been reported to cause large shifts in the Tg of the host polymer. This behavior has been explained,in general, by the altered mobility of polymer chains at an interface [163,182,183].

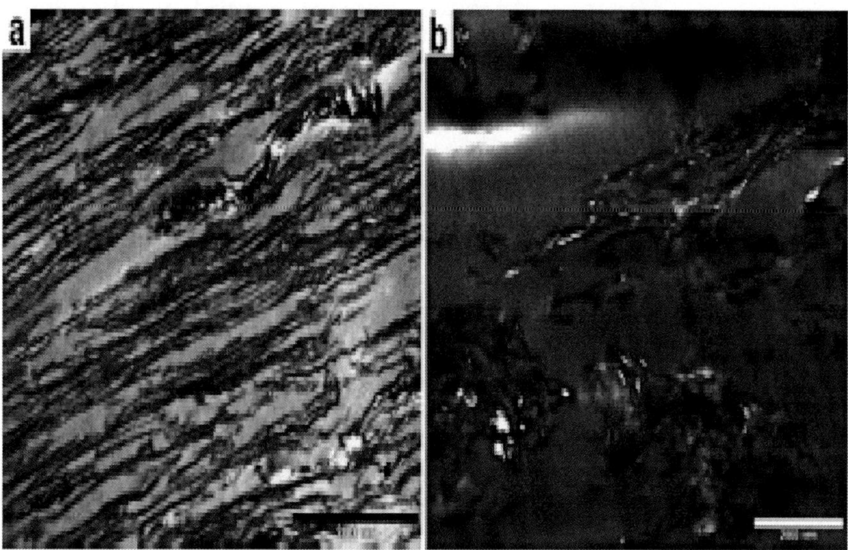

Figure 8. TEM images contrasting (a) the preferential orientation of RGeO platelets parallel to the surface of a solution-cast RGeO/Nafion and (b) the randomly oriented dispersion of TEGO platelets in Nafion (scale bars ¼ 2 mm; adapted from Ref. [162]).

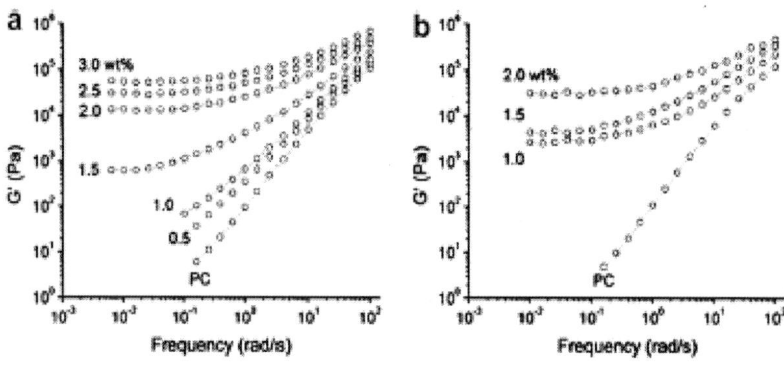

Figure 9. Dynamic frequency sweeps of melt-blended TEGO/polycarbonate composite melts, illustrating the changes in low-frequency moduli of the composites after annealing times of (a) 10,000 s and (b) 20,000 s (adapted from Ref. [153]).

Fundamentally, an attractive polymerematrix interface could restrict the chain mobility and thus tend to raise the Tg, whereas free surfaces and repulsive interfaces enhance chain mobility and lower the Tg; these mobility effects have been found to propagate away from the interface and gradually taper off with distance [184]. Depending on the strength of the interaction between the polymer and filler, this 'interphase' region of polymer chains with altered mobility may extend tens or even hundreds of nanometers away from the interface [185], potentially creating an enormous volume of polymer with significantly altered viscoelastic behavior. Formation of a network of interphase polymer may thus manifest large increases in the nanocomposite Tg at low loadings [186].

A striking example of this behavior was reported in nanocomposites of TEGO and poly(acrylonitrile) (PAN), where a shift in Tg of 40 C at only 0.05 wt% loading of TEGO was observed [7]. In addition to the residual hydroxyl groups present on the TEGO surfaces promoting a positive Tg shift due to favorable non-covalent interactions with the polymer, the nanoscale roughness of the TEGO sheets was suggested to accentuate the effect [7]. Tg shifts have also been observed for CMG/polymer nanocomposites where the polymer is covalently bound to the platelet surface.

In one report, GeO with PVA chains grafted to the surface via esterification was incorporated into a bulk PVA matrix, and the resulting composites showed a 35 C shift in Tg [187]. Other reports on CMG composites with covalent matrixefiller interfaces have shown smaller but significant Tg shifts [96,188]; for such composites, it has been reported that higher grafting densities and lower molecular weight of grafted chains correlated with higher Tg values [130]. In general, Tg shifts over 20 C are unusualdsome studies have reported maximum Tg shifts between 10 and 20 C [165,175,189], but many are lower still. Decreases in Tg have also been noted in composites which otherwise showed improvements in stiffness and electrical conductivity [157].

ELECTRICAL PERCOLATION AND CONDUCTIVITY

One of the most promising aspects of graphene-based materials is their potential for use in device and other electronics applications, owing to their high electrical conductivity. 'Paper' materials made of stacked RGeO platelets have been reported to exhibit conductivities as high as 35,100 S/m [190], and such highly conductive materials, when used as fillers, may increase the bulk conductivity of an otherwise insulating polymer (e.g., poly(styrene) [55], poly(ethylene terephthalate) [113], etc.) by several orders of magnitude. In order for a nanocomposite with an insulating matrix to be electrically conductive, the concentration of the conducting filler must be above the electrical percolation threshold, where a conductive network of filler particles is formed [191]. As shown in Fig. 10, once electrical percolation has been achieved, the increase in conductivity as a function of filler loading can be modeled by a simple power-law expression:

$$\sigma_C = \sigma_f(\phi - \phi_c)^t$$

where f is the filler volume fraction, fc is the percolation threshold, nsf is the filler conductivity, s is the composite conductivity, and t is a scaling exponent. The filler need not be in direct contact for current flow; rather, conduction can take place via tunneling between thin polymer layers surrounding the filler particles, and this tunneling resistance is said to be the limiting factor in the composite conductivity [192,193]. Interestingly, recent work on TEGO/poly(vinylidene fluoride) (PVDF) nanocomposites showed a decrease in composite resistivity with increasing temperature (a negative temperature coefficient, or NTC, effect), which may suggest that for this system that the interplatelet contact resistance dominates over the tunneling resistance [177].

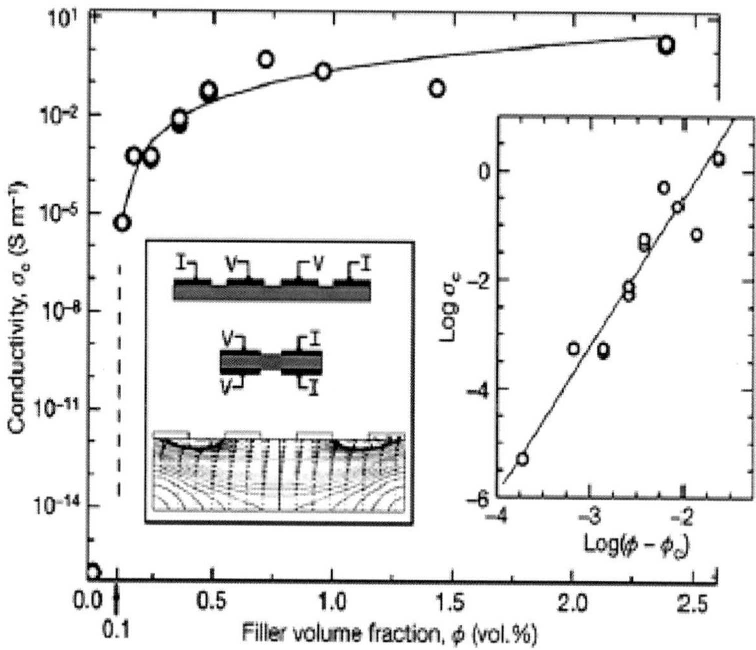

Figure 10. Conductivity of composites of PS filled with phenyl isocyanate-functionalized RG-O versus filler volume fraction, illustrating the power-law dependence of conductivity above the percolation threshold fc (adapted from Ref. [55]).

The electrical percolation thresholds achieved with graphenebased nanocomposites are often compared with those reported for CNT/polymer composites. Comparison of the sampling of percolation thresholds shown in Table 1 (representing the lowest threshold values reported to date) with comprehensive data compiled on CNT-based nanocomposites [194] reveals that the lowest values reported for GNP- and graphene-based nanocomposites are, in general, somewhat higher than those reported for CNT-based composites. In particular, CNT/epoxy nanocomposites have been reported with electrical percolation thresholds as low as approximately 0.0025 wt% [195,196], far lower than has been reported for any graphene-based nanocomposite. These exceptional results have been ascribed to 'kinetic percolation' that most notably arises in composites with a low viscosity during processing (e.g., pre-cured epoxy), which can induce formation of a flocculated network of CNTs that electrically percolates at a much lower loading than possible with well-dispersed, randomly oriented fillers [156,197]. While comparison of the electrical percolation thresholds of composites is often valid, one must consider the influence of the sample geometry used for the conductivity measurements. For instance, a sufficiently long nanotube may be able to bridge between the electrodes for a sufficiently small test specimen, potentially suggesting a low percolation threshold which would otherwise not exist in a bulk composite specimen of much larger size.

Table 1. Values of the lowest electrical percolation thresholds and maximum electrical conductivities which have been reported in the literature for GNP and graphene-based nanocomposites for selected polymer matrices

Matrix polymer	Filler type	Lowest percolation threshold reported (wt%)	Ref.	Filler type	Maximum conductivity (S/cm)[a]	Ref.
Epoxy	Funct. EG	1.0	[74]	RG-O	~0.05 (19 wt%)	[208]
Nylon-6	GO	0.5	[168]	GO	8.4×10^{-3} (1.8 wt%)	[168]
Poly(aniline) (doped)	GNP	0.7	[298]	GNP	522 (10 wt%)	[298]
Polycarbonate	TEGO	0.3	[157]	TEGO	0.5 (4.8 wt%)	[157]
Poly(ethylene)	RG-O	0.2	[198]	RG-O	0.1 (1.3 wt%)	[198]
Poly(ethylene terephthalate)	TEGO	1.0	[113]	TEGO	0.02 (6.5 wt%)	[113]
Poly(methyl methacrylate)	GNP	0.7	[84]	GNP	~1 (10 wt%)	[84]
Poly(propylene)	GNP	0.7	[119]	GNP	5×10^{-3} (10 wt%)	[119]
Poly(styrene)	Funct. G-O	0.2	[55]	RG-O	0.15 (2 wt%)	[147]
Poly(vinyl alcohol)	RG-O	0.5	[165]	RG-O	0.1 (7.5 wt%)	[165]
Poly(vinyl chloride)	GNP	1.4	[299]	GNP	0.06 (14.8 wt%)	[299]
Poly(vinylidene fluoride)	TEGO	2.0	[177]	TEGO	3×10^{-4} (4 wt%)	[177]
Polyurethane[b]	TEGO	0.6	[26]	TEGO	N/A (3.6 wt%)	[26]

[a] When loading was reported in volume percent, the density of bulk graphite (2.2 g/cm3) was used to convert to a weight percent loading.

[b] Minimum resistance reported: w200 U.

It has been said that a high degree of dispersion may not necessarily yield the lowest onset of electrical percolation [156], as a sheath of polymer may coat the surfaces of well-dispersed filler and prevent direct interparticle contact. Indeed, the lowest percolation threshold achieved thus far for a graphene-based polymer nanocomposite (approximately 0.15 wt%; see Table 1)was observed when the filler was not homogeneously dispersed in the polymer matrix, but rather segregated from the matrix to form a conductive network [198]. In this study, poly(ethylene) particles were mixed with GeO in a water/ethanol mixture and were reduced using hydrazine, causing agglomeration of the RGeO and subsequent deposition onto the poly(ethylene) particles. This heterogeneous system was then hot pressed to generate a composite with a segregated, highly conducting network of RGeO filler [198]; however, such a morphology could compromise the composite's mechanical properties due to the agglomeration of filler [24]. In a related approach, an emulsion mixing method was used to coat polycarbonate microspheres with TEGO prior to compression molding which lowered the percolation threshold by over 50% versus a standard solution mixing method (to approximately 0.31 wt%, from 0.84 wt%) [157]. TEM observations showed a uniform dispersion of TEGO in the solution-mixed composites, compared with a segregated conductive network of TEGO in the emulsion-mixed composites, perhaps due to the exclusion of TEGO from the microspheres.

Moreover, these composites (made by both dispersion methods) showed improved mechanical properties [157]. Alignment of the filler also plays a major role in the onset of electrical percolation: when the platelets are aligned in the matrix, there are, at least at relatively low concentrations, fewer contacts between them, and thus the percolation threshold would be expected to increase [199]. Compression molded polycarbonate and TEGO/polyester composites with aligned platelets were reported to show an electrical percolation threshold roughly twice that of annealed samples with randomly oriented platelets [148,153], while in another study, injection molding was reported to raise the percolation threshold over an order of magnitude versus compression molding for a GNP/poly(propylene) composite [119]. In general, the percolation threshold for electrical conductivity is often slightly higher than for rheological percolation, due to the requirement for closer proximity between platelets for particle tunneling (approximately 5 nm) for electrical percolation versus bridging by the interphase, which may extend over tens of nanometers [32,185,200]. In addition to lowering the percolation threshold, slight aggregation of the conductive filler may also improve the maximum electrical conductivities of these composites [194,201]. A combination of conductive carbon fillers may also be beneficial for lowering the electrical percolation threshold of graphene-based nanocomposites [202].

The electrical percolation threshold also depends on the intrinsic filler properties, and both theoretical models [203,204] and experiments [205] suggest that the electrical conductivity of a CMG/polymer nanocomposite depends strongly on the aspect ratio of the platelets, with a higher aspect ratio translating to a higher conductivity. Transistors produced using a phenyl isocyanate-functionalized RGeO/PS composite as the active layer reportedly exhibited an increased carrier mobility for composites containing larger-area platelets, suggesting that sheetesheet junctions limit the composite conductivity [205] (Fig.11).Wrinkled, folded, or otherwise non-ideal platelet conformations may also raise the electrical percolation threshold [206].

Aside from being used to impart electrical conductivity to an insulating polymer host, graphene-based fillers can also endow other unique electrical properties to composites. The positive temperature coefficient of resistivity of an RGeO/poly(ethylene) composite was stated to be tunable by varying the time of an isothermal heat treatment of the composite (at 180 C), which was thought to randomize the RGeO network and raise the resistivity, thus raising the PTC [207]. Graphene-based composites are being explored for their dielectric properties [208], and in one study a large increase in the dielectric constant (up to 4.5 <?> 107 at 1000 Hz) was reported in a GNP/poly(vinylidene fluoride) composite near the percolation threshold of the composite [209].

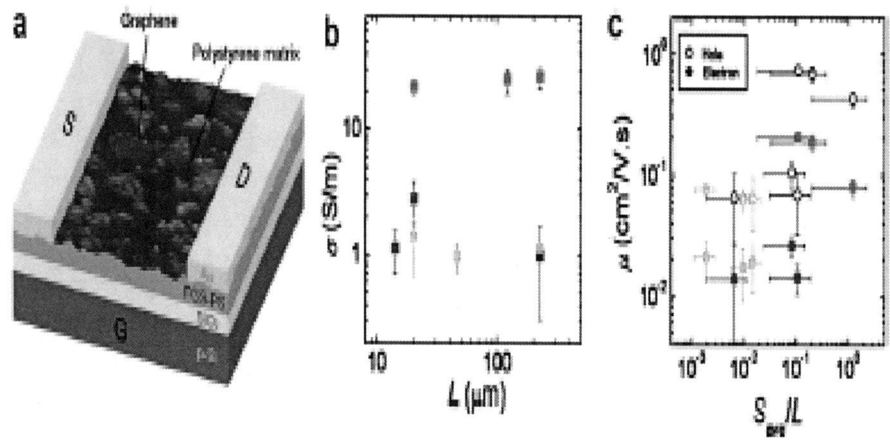

Figure 11. (a) A transistor based off a phenyl isocyanate-functionalized RGeO/poly(styrene) composite shows different levels of (b) electrical conductivity and (c) carrier mobility depending on the aspect ratio of the platelets. The average lateral dimensions for the three platelet sizes studied were 0.44 mm, 1.5 mm, and 22.4 mm for the light grey, black, and medium grey points, respectively (adapted from Ref. [205]).

Notably, a variety of device applications have also been explored for graphene-based composites and these will be discussed in a following section.

REINFORCEMENT AND MECHANICAL PROPERTIES

The in-plane elastic modulus of pristine, defect-free graphene is approximately 1.1 TPa and is the strongest material that has ever been measured on a micron length scale [210]. CMG platelets exhibit an appreciably lower in-plane stiffness which calculations suggest may scale inversely with an increasing level of oxidation of the platelets [211]. A study using AFM nanoindentation on suspended CMG platelets reported the opposite, with the elastic modulus of the platelets evidently increasing with increasing oxidation level (decreasing conductivity), ranging from 250 GPa for RGeO platelets up to approximately 650 GPa for GeO platelets [22].

In another study, AFM tip-induced deformation on suspended platelets yielded an elastic modulus of monolayer GeO platelets of approximately 208 GPa. Moreover, the measurements revealed similar modulus values for two- and three-layer GeO platelets, at low strain [21]. These relatively large modulus values (compared to most polymeric materials), coupled with the large surface areas of the platelets, allow GO-derived fillers to be the primary loadbearing component of a polymer nanocomposite [161]. Fig. 12 provides an example of the reinforcing effect reported for RGeO/ PVA composites.

Once dispersed in a polymer matrix, these compliant sheets or thin platelets commonly adopt wavy or wrinkled structures which may effectively reduce these modulus values [149], as crumpled platelets would tend to unfold rather than stretch in-plane under an applied tensile stress. Moreover, platelet restacking or incomplete exfoliation to single platelets could also lead to lower effective modulus values due to the decreased aspect ratios [161]. Highly crumpled conformations are often reported in composites using TEGO as filler, which along with structural defects generated during the high-temperature exfoliation processes [15], may significantly reduce the effective stiffness of the platelets and thus diminish their reinforcing capability. For instance, polycarbonate and poly (ethylene-2,6-naphthalate) (PEN) composites filled with TEGO showed only slightly larger modulus gains than composites reinforced with graphite at equivalent loadings, and calculations based on this experimental data suggested an effective modulus of around 70 GPa for TEGOdless than one-third of the measured value for single-layer RGeO [148,153]. Conversely, other studies have reported significant reinforcement from TEGO, attributed to strong interfacial bonding augmented by mechanical interlocking with the matrix due to the nanoscale roughness of the platelets [7,212]. Notably, the composites showing weaker reinforcement from TEGO were processed with melt mixing, which could possibly reduce the platelet aspect ratio (and thus reinforcement) due to particle attrition [32].

Aside from these issues with the intrinsic structure of GOderived fillers, the reinforcing effectiveness observed from these materials thus far may be limited by problems with interfacial adhesion and spatial distribution of filler. Mechanical property enhancements have been found to correlate with improved nanofiller dispersion [23], and alignment of the filler in the matrix may increase reinforcement

[161]. However, there is evidence that nanofillers (including CNTs and organoclays) which appear uniformly dispersed on short length scales may actually be aggregated into micron-scale fractal-like structures [156,213,214]. It has been suggested that such aggregates may be highly compliant and could reduce the effective aspect ratio of the filler, with both factors diminishing the reinforcing effect [156]. On the other hand, some have suggested that the presence of large-scale aggregates of filler is beneficial for reinforcement [138,215]. In any case, consideration of higher levels of nanocomposite structural hierarchy (as characterized by SAXS and USAXS, for example) may ultimately be necessary for extracting the full reinforcement potential of graphene-based fillers, and polymer-grafted platelets could potentially provide one route to tailoring the spatial distribution of filler [138,215].

Strong interfacial adhesion between the platelets and polymer matrix is also crucial for effective reinforcement [111,214,216e218]. Aside from making dispersion difficult, incompatibility between the phases may lower stress transfer due to low interfacial adhesion, resulting in a lower composite modulus [219]. Measurements of graphene-polymer interfacial adhesion have been carried out using AFM and Raman spectroscopy [220,221]. Evaluation of Raman spectra measured under strain revealed an interfacial shear stress of approximately 2.3 MPa in a graphene/PMMA composite (where the graphene was produced by micro-mechanical exfoliation) [220]. This value is similar to the value (2.7 MPa) of interfacial shear stress predicted by simulations of a poly(ethylene)/CNT nanocomposite with only van der Waals interactions between the matrix and filler [222], suggesting that the interaction between PMMA and the monolayer graphene platelets was also mediated by weak dispersive forces. By comparison, interfacial shear stresses up to 47 MPa have been measured for CNT/polymer composites [223] and values up to 500 MPa have been predicted for CNT-filled composites with covalent bonding at the matrixefiller interface [224]. These measurements thus suggest low levels of reinforcement for graphene-based polymer nanocomposites in the absence of covalent or stronger non-covalent bonding between the phases, emphasizing the importance of 'engineering' the fillerematrix interface in these systems.

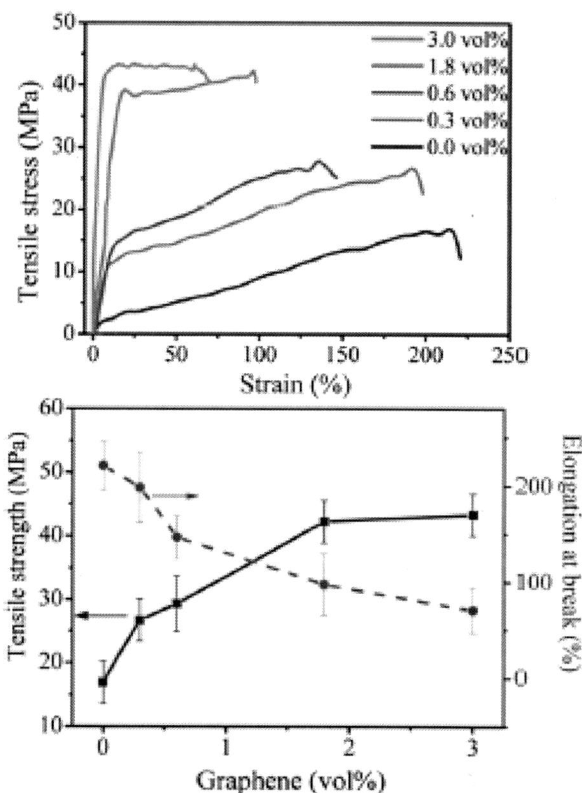

Figure 12. Stressestrain plots of RGeO/poly(vinyl alcohol) composites as a function of filler loading, showing the pronounced reinforcing effect of RGeO. Tensile strength and elongation at break show opposing trends with increasing volume fraction of RGeO (adapted from Ref. [102]).

Small molecule functionalization of graphene-based materials, either covalent or non-covalent, is a route to tailor the interface to promote stronger non-covalent interactions between the matrix and platelets [26]. Hydrogen bonding between GO-derived fillers and their polymer hosts has been cited as a major factor in large modulus and strength improvements observed in several polymers that can serve as hydrogen bond acceptors and/or donors [7,101,105,225]. Formation of a crystalline layer around the platelets by an isothermal treatment may also enhance stress transfer, as has been reported for a GeO/poly(caprolactone) composite [167].

Covalent bonding between the filler and matrix may provide the most effective means to increase the interfacial shear stress for improving stress transfer [218]. Monomers or chain extenders containing functional groups that can react directly with GeO in an in situ step-growth polymerization may thus covalently link GeO to the matrix; as recently reported for a GeO/Nylon-6 nanocomposite, this approach resulted in an unprecedented doubling of modulus and strength versus neat Nylon-6 at just 0.1 wt% loading [141]. The formation of covalent bonds between matrix and filler has also been reported to improve mechanical properties in epoxy and polyurethane composites with GNP and GO-derived fillers [26,74,140,226e228]. In addition, incorporation of polymer-grafted CMG fillers has been reported to greatly enhance the mechanical properties of PS [96], PMMA [129,188], and PVDF [128]. For such polymer-grafted fillers incorporated into a chemically identical polymer matrix, improvements in reinforcement may be strongly influenced by the relative molecular weights of the grafted polymer and matrix polymer [229].

Particularly largemodulus gains have beenreported in elastomeric matrices, most notably polyurethanes [26,140,226,227,230e232]. It has been pointed out that the pronounced modulus increases of elastomers may arise from the large difference in modulus between the filler and matrixdwhich also makes elastomers less sensitive to filler defects and non-ideal conformations than rigid thermoplasticsdindeed, for graphene-based polyurethane composites the reinforcement effectwas reported to be significantly attenuated above the soft segment Tg (approximately 30 C) [26]. Increases inmodulus of over two orders ofmagnitude (fromapproximately 10 MPa to 1.5 GPa at 55 wt% GNP [230]) have been reported in polyurethanes. In one study, moderate ductility was said to be retained despite the high loading, affording a composite with modulus and strength comparable to many rigid thermoplastics (e.g., polycarbonate) but with amuch higher toughness and strain at break (15%) [230]. As shownin Fig. 13, TEGO/polyurethane composites made via in situ polymerization showed less improvement in modulus as compared with solution-mixed composites despite good dispersion, possibly due to chain extension by TEGO which may have inhibited formation of ordered, hydrogen-bonded hard segments [26]. In the case of a TEGO/silicone foam nanocomposite, densification of the composite relative to the neat foam may have complemented the effect of particle reinforcement, leading to the observed 200% increase in the normalized compressive modulus at just 0.25 wt% [233]. While the reported modulus increases are often significant, generally the remarkable ductility of elastomers is significantly compromised by incorporation of rigid filler; furthermore, graphene-filled elastomers often show a decline in tensile strength.

Beyond simple reinforcing effects, improvements in fracture toughness, fatigue strength, and buckling resistance have been noted in graphene-based composites [158,180,212,234e236]. At equivalent loadings, in situ polymerized TEGO/epoxy composites reportedly showa much higher buckling strength, fracture strength, and fracture energy than single- or multi-walled nanotubes-filled composites. In one study, a 94% increase in the fracture toughness of an CMG/epoxy composite versus neat epoxy was reported at 0.6 wt % loading. The toughness increase was attributed to the presence of pendant amine functionality on the CMG platelets potentiallycreating a "flexible interphase," although cross-linking between the CMG platelets may have also played a role [180].

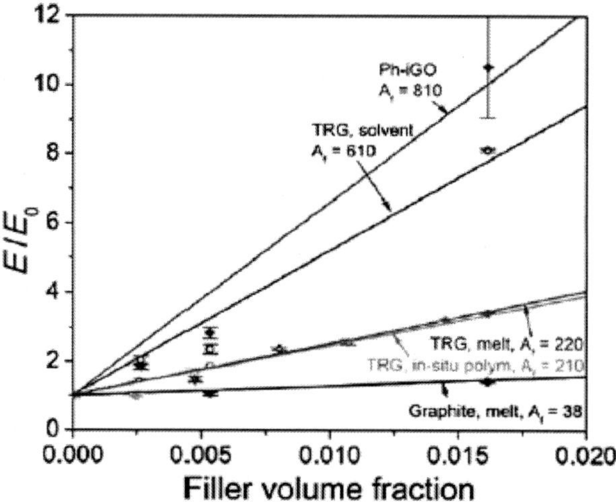

Figure 13. Elastic moduli of polyurethane composites with flake graphite, TEGO ('TRG'), and phenyl isocyanate-functionalized GeO platelets (PheiGO) as filler, processed by either melt mixing, solvent mixing, or in situ polymerization (PheiGO composites were made by solution mixing). Fitting of the data to micro-mechanical models (solid lines) allows for calculations of effective aspect ratios (Af) to quantify dispersion (adapted from Ref. [26]).

Both increases [236] and decreases [237] in the impact strength of graphene-based composites have been reported. As suggested above, GO-derived fillers could show superior reinforcement to layered clays due to their higher intrinsic stiffness. Calculations have suggested that randomly oriented graphene platelets may also produce nanocomposites with higher stiffness and strength than randomly oriented nanotubes, although calculations suggest that aligned CNTs may provide better reinforcement than aligned platelets at equivalent loading, aspect ratio, and dispersion [238]. Comparisons with GNP-filled composites [36,116,117,119,176,239] reveal that, generally, better mechanical properties are evidently achieved using well-exfoliated, GOderived fillers at equivalent loadings. This may be primarily due to the relatively large platelet thicknesses (with a stacked structure) of GNP fillers, resulting in a lower aspect ratio and lower effective platelet modulus, thus decreasing their reinforcing effect. Notably, filler combinations such as SWNTs and TEGO [240] as well as TEGO and GO [241] may provide synergistic reinforcement, although the benefit (or lack of benefit) of these systems versus a single CMG filler at the same loading has not been fully established. Comparisons of micro-mechanical predictions (e.g., Mori-Tanaka and Halpin-Tsai models) with experimental results on nanocomposites seem to indicate that reinforcement arises predominately from the native properties of the filler [111,161]. On the other hand, it has been suggested that the interphase concepts used to explain viscoelastic behavior may play a significant role in reinforcement owing to the modification of the properties of a large volume fraction of the matrix [163,216,242e244]. While these micro-mechanical models have been applied to some graphenebased composites in some reports [102,148,153], other results using GO-derived fillers have reportedly shown reinforcement at low loadings that exceed the upper-bound modulus predictions of these micro-mechanical models, perhaps suggesting a non-negligible contribution from the modified interphase matrix for such 'strongly bonded' systems [7,158]. Regardless, the apparent discrepancy between these results and theory highlights the need to develop further understanding of the relative contributions of native filler properties and changes in the polymer matrix in regards to the reinforcement of these systems.

THERMAL CONDUCTIVITY, THERMAL STABILITY, AND DIMENSIONAL STABILITY

The exceptional thermal properties of GNPs and graphenebased materials have been harnessed as fillers to improve the thermal conductivity, thermal stability, and dimensional stability of polymers. Pristine graphene is highly thermally conductive; at room temperature, the thermal conductivity has been measured to be exceed 3000 W/m K when suspended [245,246] and approximately 600 W/m K when supported on a SiO2 substrate [247]. CNTs show similar intrinsic thermal conductivities, but the sheet-like geometry of graphene-based materials may provide lower interfacial thermal resistance and thus produce larger conductivity improvements in polymer composites [248,249]. The geometry of graphite and graphene filler can also impart significant anisotropy to the thermal conductivity of the polymer composite [36], with the measured in-plane thermal conductivity as much as ten times higher than the cross-plane conductivity [250,251].

Many of the same considerations discussed for reinforcement and electrical conductivity apply for improving thermal conductivity. Close interparticle contact reduces thermal resistance and thermal conductivity in nanocomposites has been rationalized with percolation theory [252,253]. Since phonons are the primary mode of thermal conduction in polymers, covalent bonding between the matrix and filler can reduce phonon scattering at the matrixefiller interface, promoting higher composite thermal conductivity [175].

However, just as with electrical conduction, thermal conduction in nanocomposites can be compromised to some degree by platelet functionalization to enhance interfacial bonding [254].

Thermal conductivity studies of GNP and graphene-based composites have largely focused on epoxy matrix composites [175,179,248,250,255e257]. Significant improvements in thermal conductivity have been achieved in these systems (with composite conductivities ranging from 3 to 6 W/m K, up from approximately 0.2 W/m K for neat epoxy), but such large gains require relatively high carbon loadings (20 wt% and higher). Conductivities as high 80 W/m K at 64 wt% have been reported for GNP/epoxy composites, based on measurements of the thermal diffusivity [250]. The much smaller thermal conductivity contrast between polymer matrices and carbon nanofillers (as compared with electrical conductivity) may be the reason for the lower increases in thermal conductivity observed versus electrical conductivity at a given loading [95].

Different techniques have been utilized to lower the filler loading necessary to achieve large thermal conductivity gains. For instance, functionalization of EG with amine silyl groups improved the thermal conductivity by up to 20% over unmodified EG at the same loading [175]. A synergistic effect of combining SWNTs and GNPs (maximizing at a ratio of approximately 3:1 of GNPs:SWNTs) was reported, with TEM observations by the authors used to attribute the effect to the morphology of the filler blend in which the SWNTs bridged across adjacent GNP platelets, forming an extended network of filler in direct contact (Fig. 14) [248]. For GeO platelets with surface-initiated polymer brushes, higher composite thermal conductivities were reported to correlate with lower molecular weights of polymer and lower initiator grafting densities, perhaps illustrating the negative effect of excessive functionalization on thermal conduction [130].

A significant number of reports have reported increased thermal stability (as typically defined by the maximum mass loss rate measured by thermogravimetric methods) of polymers using GNPs and various CMGs as filler [114,116,170,177]; even GO can enhance the overall composite thermal stability versus the neat polymer [141,181], despite being thermally unstable itself. Studies on nanoclay-filled composites have suggested improved thermal stability trends with increased levels of exfoliation and interfacial adhesion [258]. Most of these studies have focused on non-oxidative stability (heating under inert gas such as nitrogen or argon) and comparatively little data exists on the oxidative stability of GO composites. Increases in the onset of (non-oxidative) degradation of 20e30 C and higher have been reported with GO-derived fillers [7,103,104,177]. However, GO is not always observed to confer thermal stabilitydin one case the presence of GO filler was found to accelerate the decomposition kinetics [259].

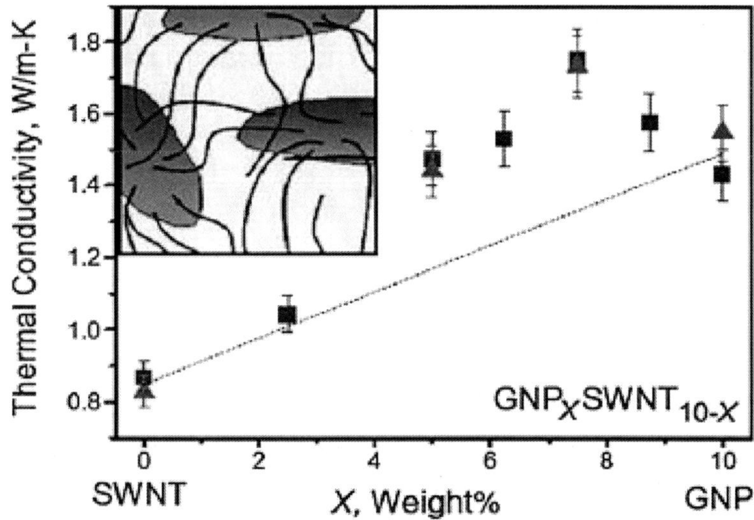

Figure 14. Combination of graphite nanoplatelets (GNP) and single-walled carbon nanotubes (SWNTs) synergistically improve the thermal conductivity of epoxy. TEM studies established the presence of SWNTs bridging between dispersed GNPs (as shown in the schematic, inset) which may be responsible for the effect (adapted from Ref. [248]).

The negative coefficient of thermal expansion (CTE) of graphene [260,261], along with its high specific surface area and high stiffness, can significantly lower the coefficient of thermal expansion (CTE) of a polymer matrix [111]. In one study, the CTE of a GeO/epoxy composite was lowered by nearly 32% for temperatures below the Tg of the matrix at 5 wt% loading, although at 1 wt%, SWNTs produced a larger decrease in CTE than GeO platelets also loaded at 1 wt% (the two fillers were not compared at 5 wt%) [255]. Compared with CNTs, GNPs were reported to decrease the CTE of PP in two directions instead of one when aligned in the matrix [36].

GAS BARRIER PROPERTIES

The incorporation of GNPs and GO-derived fillers can significantly reduce gas permeation through a polymer composite relative to the neat matrix polymer. A percolating network of platelets can provide a 'tortuous path' which inhibits molecular diffusion through the matrix, thus resulting in significantly reduced permeability (Fig. 15) [111]. Permeability studies on CMG/PS nanocomposites, however, suggest that at low loadings (e.g., below 0.05 vol%), the reduction in permeability of the composite is largely driven by a reduction in gas solubility in the composite, with diffusion effects becoming more important at higher loadings [262]. Orientation of the platelets may further enhance barrier properties perpendicular to their alignment, while higher platelet aspect ratios correlate with increased barrier resistance [111].

Notably, the one-dimensional geometry of CNTs may limit their effectiveness in improving the barrier properties of a composite relative to the neat polymer [26]. Both GNP and GO-derived fillers have been investigated in various permeation studies [26,36,114,115,153,262]; for thermoplastics, results include a 20% reduction in oxygen permeability for PP with 6.5 wt% GNP [36], and a 39% reduction in the nitrogen permeability of a polycarbonate/TEGO composite at approximately 3.5 wt% loading [153]. As shown in Fig.15, CMG/PS composites were reported to show a lower oxygen permeability than exfoliated montmorillonite/PS composites at equivalent loadings [262]. In a comparative study of GO-derived fillers in thermoplastic polyurethanes, phenyl isocyanate-functionalized GeO platelets were reported to confer superior barrier properties relative to TEGO, with up to a 99% reduction in nitrogen permeability observed at approximately 3.7 wt% loading compared with an 81% reduction for TEGO at the same loading. Notably, the barrier properties in this study correlated with modulus improvements suggesting better filler alignment or higher aspect ratio for the functionalized GeO composites [26].

APPLICATIONS OF GRAPHENE-BASED POLYMER NANOCOMPOSITES

Though numerous challenges remain in developing a fundamental understanding of GO-derived materials and their polymer composites, these materials have already been explored for a range of applications. Reflective of GeO's close relationship to graphene, many of the applications have focused on harnessing its electronic properties, particularly for devices. However, this is somewhat counterintuitive given that GeO is essentially an insulating material. For device applications, typically the GeO platelets and polymer are mixed and then the GeO platelets are reduced using hydrazine (or other strong chemical reductants) or thermal annealing, as discussed in the examples above.

One notable exception to this paradigm can be found in the development of an electronic memory device. Using the previously described surface-initiated polymerization methods, poly(t-butylacrylate) was reportedly grown from the surface of a thin film of GeO platelets [132]. The functionalized sheets were blended with P3HT and spun cast onto a surface of indium tin oxide (ITO), and then coated with a layer of aluminum. This device made use of the low conductivity of stacked and overlapped GeO platelets as a means of creating a barrier for inter-lamellar electron-hole recombination after excitons were formed. This created a clearly defined OFF state for the memory device. At the switching potential (found to be relatively high; 1.6 V), electrons were said to be excited from the HOMO of the P3HT portion of the mixture into the LUMO of the GeO platelet film/poly(t-butylacrylate) portion via intermolecular charge transfer. As the oxygen-containing functional groups of GeO platelets are believed to be inhomogeneously distributed across the lamellae [3], the excited electrons were able to travel relatively freely within the highly delocalized p-domains.

Notably, the low bulk conductivity prevented recombination, resulting in a non-volatile memory storage system. This unique example took advantage of what is often perceived as a negative trait of GeO platelet material: its low conductivity. In contrastmetal- and dopant-free conductive composites can be prepared from electrically conductive CMGs. These studies typically utilize the high degree of surface functionalization present on GeO platelets to attach polymers or polymerization initiators (graftingto or grafting-from approaches, as previously discussed), followed by reduction of the oxidized surface to render the composite electrically conductive.

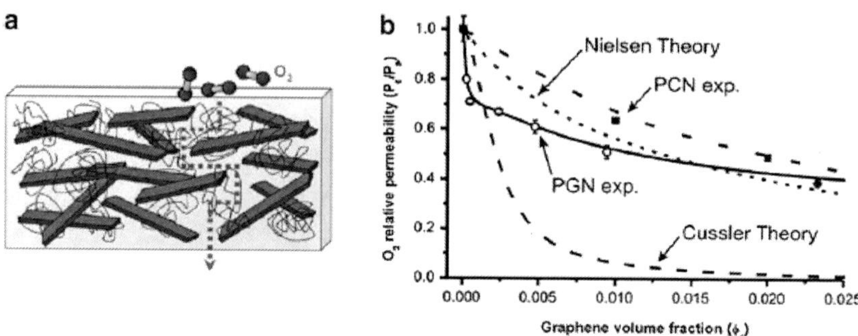

Figure 15. (a) Illustration of formation of a 'tortuous path' of platelets inhibiting diffusion of gases through a polymer composite (Nielsen model). (b) Measurements of oxygen permeability of CMG/PS ('PGN') and montmorillonite/PS ('PCN') composites as a function of filler loading, compared with two theoretical models of composite permeability (adapted from Ref. [262]).

Field effect transistors (FETs) [205], solar cells (and other optoelectronic applications) [263], and energy storage devices [264] are three areas where such conductive composites may be particularly applicable. All of these applications capitalize upon the high conductivity inherent to many CMG composites. Photovoltaics and optoelectronics, in particular, rely on the fact that monolayers of graphene are about 98% transparent but still have high electrical conductivity. This feature makes graphene-based materials potentially well-suited to address issues related to photoexcitation and exciton mobility/diffusion as transparent conducting electrodes [265e268]. Overall efficiencies for these devices are still relatively lowto date (typically less than 1%power conversion efficiency)when applied as organic photovoltaics, but this remains an area of interest particularly since the testing of these materials started so recently.

In addition to conductive polymers such as P3HT, graphenebased composites incorporating PANI have been studied as energy storage materials. Specific capacitances reported from 210 F/g to over 1000 F/g have been reported for these composites [269e274].

The precise reasons for this enhancement in energy storage capacity are still under investigation, but the authors state that this likely involves Faradaic transitions between the three different oxidation states of PANI (leucoemeraldine, emeraldine salt, and pernigraniline), in addition to the electrochemical double-layer capacitance (EDLC; general schematic shown in Fig.16) provided by the carbon material [275]. It has been reported that PANI can intercalate into the layers of unexfoliated GO, increasing compatibility between the phases [181,276e278]. Even when GO was not reduced, the composite was rendered conductive by the PANI [276,277], though to a lesser extent than when RGeO platelets were incorporated [269].

Aside from devices, a host of diverse applications have been envisioned for graphene-based nanocomposites, all harnessing the property improvements discussed in the previous sections. For instance, the electrical conductivity of these composites may find use in electromagnetic wave interference shielding and anti-static coatings [208], while potentially maintaining properties of the host polymer such as transparency by virtue of the low percolation thresholds of these composite systems [55]. The combination of the improved barrier properties and increased light absorption of a CMG/PS composite versus neat PS suggests wider application as a packaging material [262]. The mechanical reinforcement achieved at low loadings of GO-derived filler offers potential uses in weight-sensitive aerospace and automotive applications such as tires, which could also benefit from the conductivity of RGeO platelets. Electrically conductive and robust GO-derived composite membranes could find use as capacitive pressure sensors in MEMS applications [279]. One less conventional application for these composites may be in self-healing materials [280,281]: very small loadings of multi-layer graphene in a shape memory epoxy matrix was reported to improve the composite's resistance to crack formation, thus enhancing scratch recovery upon heating above its Tg [282]. Interestingly, the filler used in this study was grown via microwave plasma-enhanced chemical vapor deposition, representing the first report of graphene composite filler generated from a 'bottom-up' synthesis route.

Nanofillers can also be used to reduce or overcome the intrinsic flammability of thermoplastics [111], and such applications have been explored for GO, despite the high flammability of GO when contaminated by synthetic byproducts such as potassium salts [283]. However, the thermal expansion of GO coupled with its evolution of gaseous byproducts (e.g., CO2) at high temperatures may confer flame retardancy [284]. A recent study compared the fire retardancy of a GO/Nylon-6 composite to a nanoclay/Nylon-6 composite, and concluded that the volumetric expansion of GO particles conferred good short-term fire resistance, but the low structural integrity of the resulting material and release of oxidants on heating was a major disadvantage relative to nanoclays [285].

An emerging research direction for graphene-based composites is focused on biomedical applications. Graphene has been investigated for biosensor applications [286] and efforts have been directed at graphene-based composite biosensors as well [287e289]. Composite films of DNA and RGeO platelets have been prepared via solution mixing, and incorporation of other biological macromolecules along with DNA may provide a general approach to multifunctional, biocompatible composites [290]. A biocompatible hydrogel composite produced via physical cross-linking of PVA chains between GeO platelets showed a controlled release of Vitamin B12 depending on solution pH, and the authors stated this could find use in drug delivery [291]. Moreover, biocompatible freestanding composite films of poly(oxyethylene sorbitan laurate) (TWEEN) and RGeO could find use in transplant devices and implants [292]. Other biocompatible and biodegradable polymer composites have also been investigated, and incorporation of GeO platelets into chitosan [189,225] and poly(lactide) [114,164], in particular, may greatly expand the utility of these polymers that are otherwise limited by their mechanical properties.

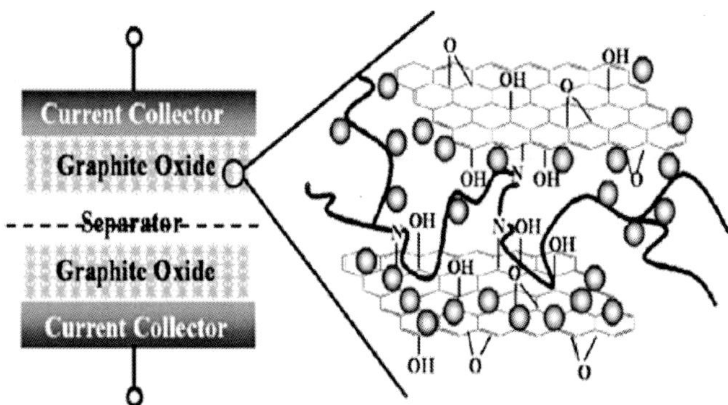

Figure 16. Schematic of test electrochemical double-layer capacitor (EDLC) assembly and structural model of a GO/polymer composite assembled in the EDLC. The spheres in represent the electrolyte ions, which can be conveyed via diffusion and the segmental motion of the polymer chains (adapted from Ref. [264]).

CONCLUSIONS

Graphene-based polymer nanocomposites represent one of the most technologically promising developments to emerge from the interface of graphene-based materials and polymer materials. However, there are still many challenges that must be addressed for these nanocomposites to reach their full potential. For example, the sonication and thermal shock techniques commonly used to exfoliate GO also reduce the aspect ratio of the exfoliated platelets, which may negatively affect reinforcement as well as electrical and thermal properties [8,52]. Results suggesting poor interfacial adhesion in graphene/polymer composites in the absence of covalent bonding or additional non-covalent binding interactions such as pep interactions or hydrogen bonding underscore the importance of the platelet surface chemistry in reinforcement and the need for continued progress in this direction [220]. Moreover, the defects introduced into GeO platelets by either the oxidation to convert graphite to GO or the processing to generate GeO platelets, might ultimately limit the electrical conductivity and mechanical properties achievable with RGeO platelets relative to pristine and defect-free graphene platelets. Thus, methods of graphene platelet production which preserve its extended, conjugated structure may find favor for certain demanding applications of graphene-based composites [12].

Further property improvements in graphene-based composites will be influenced by improved morphological control. Defects and wrinkles in platelets are likely to influence their reinforcing capabilities, and so exfoliation and/or dispersion techniques that promote a more elongated morphology could conceivably further improve mechanical properties of these composites. Also, increased control over alignment and spatial organization of graphene-based fillers could be beneficial to nearly all types of composite properties.

Thermal annealing of graphene-based composites to randomize GNP and graphene-based platelets may benefit the composite's electrical conductivity [153], whereas improved alignment of the platelets may improve reinforcement [161]. A variety of techniques for alignment of CNTs in polymer composites have been reported [293], and some of these techniques may find use for graphene-based composites. While the end application of graphene-based composites may dictate their specific morphological characteristics, the use of 'top-down' patterning or 'bottomup' mesophase self-assembly approaches that have better morphological control could help to guide future studies of these systems, which may reveal new applications for these composites [213].

Despite these challenges, some of which are not unique to graphene-based nanocomposites, polymer nanocomposites have already found use in industry and their commercial impact is expected by many to rise significantly in the future [294]. As further improvements are made in the chemical production of graphenebased materials [295], composites using this class of filler could become a commercial reality. Notably, there is already significant effort and industrial interest in scaling up GO production [296]; GNP fillers have attracted similar attention [297]. Particularly relevant to large-scale production and transport of GO, flammability issues with this material have been identified and methods to alleviate these problems have been demonstrated [283]. Graphite, the precursor to GO and GO-derived fillers, is relatively cheap and abundant [296], and this cost factor will likely remain one of the primary advantages for using graphene-based fillers over CNTs in nanocomposites particularly as work on scale-up progresses.

Furthermore, while CNTs and graphene reportedly offer comparable mechanical and electrical property enhancements, graphene-based materials appear to provide larger thermal conductivity enhancement as well as the advantage of improving barrier properties. The multifunctional property enhancements already demonstrated with graphene-based fillers, coupled with their potential for low cost and large-scale production, may expedite the applications of these nanocomposites as well as their transition to the marketplace.

ACKNOWLEDGEMENTS

This work was supported (in part) by the Laboratory Directed Research and Development (LDRD) program and the National Institute for Nano-Engineering at Sandia National Laboratories.

Sandia is a multiprogram laboratory operated by Sandia Corporation, a Lockheed Martin Company, for the United States

Department of Energy's National Nuclear Security Administration under Contract DE-AC04-94AL85000. CWB is grateful to the Robert A. Welch Foundation (F-1621) for support.

REFERENCES

1. Zhu Y, Murali S, Cai W, Li X, Suk JW, Potts JR, et al. Adv Mater 2010;22:3906e24.
2. Geim AK, Novoselov KS. Nat Mater 2007;6:183e91.
3. Compton OC, Nguyen SBT. Small 2010;6:711e23.
4. Carter LW, Hendricks JG, Bolley DS. 2531396, National Lead Company; 1950.
5. Usuki A, Kojima Y, Kawasumi M, Okada A, Fukushima Y, Kurauchi T, et al. J Mater Res 1993;8:1179e84.
6. Winey KI, Vaia RA. MRS Bull 2007;32:314e9.
7. Ramanathan T, Abdala AA, Stankovich S, Dikin DA, Herrera-Alonso M, Piner RD, et al. Nat Nanotechnol 2008;3:327e31.
8. Dreyer DR, Park S, Bielawski CW, Ruoff RS. Chem Soc Rev 2010;39:228e40.
9. Kim H, Abdala AA, Macosko CW. Macromolecules 2010;43:6515e30.
10. Dreyer DR, Ruoff RS, Bielawski CW. Angew Chem Int Ed 2010;49:9336e44.
11. Park S, Ruoff RS. Nat Nanotechnol 2009;5:217e24.
12. Dato A, Radmilovic V, Lee Z, Phillips J, Frenklach M. Nano Lett 2008;8:2012e6.
13. Schafhaeutl C. J Prakt Chem 1840;21:129.
14. Schafhaeutl C. Philos Mag 1840;16:570.
15. Boehm HP, Clauss A, Fischer G, Hofmann U. Proceedings of the fifth conference on carbon; 1962.
16. Boehm HP, Clauss A, Fischer GO, Hofmann U. Z Naturforsch 1962;17b:150e3.
17. Boehm HP, Clauss A, Fischer GO, Hofmann U. Z Anorg Allg Chem 1962; 316:119e27.
18. Allen MJ, Tung VC, Kaner RB. Chem Rev 2010;110:132e45.
19. Loh KP, Bao Q, Ang PK, Yang J. J Mater Chem 2010;20:2277e89.
20. Rafiee MA, Liu W, Thomas AV, Zandiatashbar A, Rafiee J, Tour JM, et al. ACS Nano; 2010. doi:10.1021/nn102529n.
21. Suk JW, Piner RD, An J, Ruoff RS. ACS Nano; 2010. doi:10.1021/nl101902k.
22. Gómez-Navarro C, Burghard M, Kern K. Nano Lett 2008;8:2045e9.
23. Alexandre M, Dubois P. Mater Sci Eng R Rep 2000;28:1e63.
24. Thostenson ET, Li CY, Chou TW. Compos Sci Technol 2005;65:491e516.
25. Li D, Kaner RB. Science 2008;320:1170e1.
26. Kim H, Miura Y, Macosko CW. Chem Mater 2010;22:3441e50.
27. Dresselhaus MS, Dresselhaus G. Adv Phys 1981;30:139e326.
28. Jang BZ, Zhamu A. J Mater Sci 2008;43:5092e101.

29. Viculis LM, Mack JJ, Mayer OM, Hahn HT, Kaner RB. J Mater Chem 2005; 15:974e8.
30. Chen G, Wu D, Weng W, Cuiling W. Carbon 2003;41:619e25.
31. Li X, Zhang G, Bai X, Sun X, Wang X, Wang E, et al. Nat Nanotechnol 2008;3:538e42.
32. Potschke P, Abdel-Goad M, Pegel S, Jehnichen D, Mark JE, Zhou DH, et al. J Macromol Sci Part A Pure Appl Chem 2010;47:12e9.
33. Yasmin A, Luo J-J, Daniel IM. Compos Sci Technol 2006;66:1182e9.
34. Zheng W, Lu X, Wong S-C. J Appl Polym Sci 2004;91:2781e8.
35. Celzard A, Mareche JF, Furdin G, Puricelli S. J Phys D Appl Phys 2000; 33:3094e101.
36. Kalaitzidou K, Fukushima H, Drzal LT. Carbon 2007;45:1446e52.
37. Shioyama H. Synth Met 2000;114:1e15.
38. Novoselov KS, Geim AK, Morozov SV, Jiang D, Zhang Y, Dubonos SV, et al. Science 2004;306:666e9.
39. Zhang M, Parajuli RR, Mastrogiovanni D, Dai B, Lo P, Cheung W, et al. Small 2010;6:1100e7.
40. Hernandez Y, Nicolosi V, Lotya M, Blighe FM, Sun Z, De S, et al. Nat Nanotechnol 2008;3:563e8.
41. Lotya M, Hernandez Y, King PJ, Smith RJ, Nicolosi V, Karlsson LS, et al. J Am Chem Soc 2009;131:3611e20.
42. Lu J, Yang J-X, Wang J, Lim A, Wang S, Loh KP. ACS Nano 2009;3:2367e75.
43. Buchsteiner A, Lerf A, Pieper Jr A. J Phys Chem B 2006;110:22328e38.
44. Dikin DA, Stankovich S, Zimney EJ, Piner RD, Dommett GHB, Evmenenko G, et al. Nature 2007;448:457e60.
45. Ruess VG, Vogt F. Monatsh Chem 1948;78:222e42.
46. Park S, An JH, Jung IW, Piner RD, An SJ, Li XS, et al. Nano Lett 2009;9:1593e7.
47. Becerril HA, Mao J, Liu Z, Stoltenberg RM, Bao Z, Chen Y. ACS Nano 2008; 2:463e70.
48. Gómez-Navarro C, Weitz RT, Bittner AM, Scolari M, Mews A, Burghard M, et al. Nano Lett 2007;7:3499e503.
49. Li Z, Zhang W, Luo Y, Yang J, Hou JG. J Am Chem Soc 2009;131:6320e1.
50. Paredes JI, Villar-Rodil S, Martinez-Alonso A, Tascon JMD. Langmuir 2008;24:10560e4.
51. Zhu Y, Stoller MD, Cai W, Velamakanni A, Piner RD, Chen D, et al. ACS Nano 2010;4:1227e33.
52. Schniepp HC, Li JL, McAllister MJ, Sai H, Herrera-Alonso M, Adamson DH, et al. J Phys Chem B 2006;110:8535e9.
53. McAllister MJ, Li JL, Adamson DH, Schniepp HC, Abdala AA, Liu J, et al. Chem Mater 2007;19:4396e404.
54. Brunauer S, Emmett PH, Teller E. J Am Chem Soc 1938;60:309e19.

55. Stankovich S, Dikin DA, Dommett GHB, Kohlhaas KM, Zimney EJ, Stach EA, et al. Nature 2006;442:282e6.
56. Zhu YW, Murali S, Stoller MD, Velamakanni A, Piner RD, Ruoff RS. Carbon 2010;48:2118e22.
57. Bagri A, Mattevi C, Acik M, Chabal YJ, Chhowalla M, Shenoy VB. Nat Chem 2010;2:581e7.
58. Boukhvalov DW, Katsnelson MI. J Am Chem Soc 2008;130:10697e701.
59. Dreyer DR, Jia HP, Bielawski CW. Angew Chem Int Ed 2010;49:6813e6.
60. Jia HP, Dreyer DR, Bielawski CW. Adv Synth Catal; 2011. doi:10.1039/adsc.201000748.
61. Murali S, Dreyer DR, Zhu Y, Ruoff RS, Bielawski CW. J Mater Chem; 2011. doi:10.1039/C0JM02704A.
62. Gao J, Liu F, Liu Y, Ma N, Wang Z, Zhang X. Chem Mater 2010;22:2213e8.
63. Zhang J, Yang H, Shen G, Cheng P, Zhang J, Guo S. Chem Commun 2010; 46:1112e4.
64. Stankovich S, Dikin DA, Piner RD, Kohlhaas KA, Kleinhammes A, Jia Y, et al. Carbon 2007;45:1558e65.
65. Li D, Muller MB, Gilje S, Kaner RB, Wallace GG. Nat Nanotechnol 2008;3:101e5.
66. Stankovich S, Piner RD, Chen X, Wu N, Nguyen SBT, Ruoff RS. J Mater Chem 2006;16:155e8.
67. Choi EY, Han TH, Hong J, Kim JE, Lee SH, Kim HW, et al. J Mater Chem 2010; 20:1907e12.
68. Cao Y, Feng J, Wu P. Carbon 2010;48:3834e9.
69. Villar-Rodil S, Paredes JI, Martinez-Alonso A, Tascon JMD. J Mater Chem 2009;19:3591e3.
70. Niyogi S, Bekyarova E, Itkis ME, McWilliams JL, Hamon MA, Haddon RC. J Am Chem Soc 2006;128:7720e1.
71. Xu Y, Liu Z, Zhang X,WangY, Tian J, Huang Y, et al. Adv Mater 2009;21:1275e9.
72. Liu Y, Zhou J, Zhang X, Liu Z, Wan X, Tian J, et al. Carbon 2009;47:3113e21.
73. Stankovich S, Piner RD, Nguyen ST, Ruoff RS. Carbon 2006;44:3342e7.
74. Miller SG, Bauer JL, Maryanski MJ, Heimann PJ, Barlow JP, Gosau JM, et al. Compos Sci Technol; 2010:1120e5.
75. Su Q, Pang S, Alijani V, Li C, Feng X, Müllen K. Adv Mater 2009;21:3191e5.
76. Xu Y, Bai H, Lu G, Li C, Shi G. J Am Chem Soc 2008;130:5856e7.
77. Podall H, Foster WE, Giraitis AP. J Org Chem 1958;30:82e5.
78. Panayotov IM, Rashkov IB. J Polym Sci Part A Polym Chem 1973;11:2615e22.
79. Parrod J, Beinert G. J Polym Sci 1961;53:99.
80. Shioyama H. Carbon 1997;35:1664e5.
81. Bunnell LR. 5186919, Battelle Memorial Institute; 1993.

82. Pan YX, Yu ZZ, Ou YC, Hu GH. J Polym Sci Part B Polym Phys 2000;38:1626e33.
83. Chen G, Zhao W,editors. Nano- and biocomposites. CRCPress; 2009. p. 79e106.
84. Chen G, Weng W, Wu D, Wu C. Eur Polym J 2003;39:2329e35.
85. Chen G, Wu C, Weng W, Wu D, Yan W. Polymer 2003;44:1781e4.
86. Zheng W, Wong S-C. Compos Sci Technol 2003;63:225e35.
87. Zheng W, Wong S-C, Sue H-J. Polymer 2002;43:6767e73.
88. Zois H, Apekis L, Omastova M. Proceedings e international symposium on electrets, 10th international symposium on electrets; 1999: pp. 529e532.
89. Gabriel P, Cipriano LG, Ana JM. Polym Compos 1999;20:804e8.
90. Liu P, Gong K. Carbon 1999;37:701e11.
91. Kyotani T, Moriyama H, Tomita A. Carbon 1997;35:1185e7.
92. Matsuo Y, Tahara K, Sugie Y. Carbon 1997;35:113e20.
93. Matsuo Y, Hatase K, Sugie Y. Chem Mater 1998;10:2266e9.
94. Kotov NA, Dékány I, Fendler JH. Adv Mater 1996;8:637e41.
95. Moniruzzaman M, Winey KI. Macromolecules 2006;39:5194e205.
96. Fang M, Wang KG, Lu HB, Yang YL, Nutt S. J Mater Chem 2009;19:7098e105.
97. Higginbotham AL, Lomeda JR, Morgan AB, Tour JM. ACS Appl Mater Interfaces 2009;1:2256e61.
98. Pandey R, Awasthi K, Tiwari RS, Srivastava ON. 2010; arXiv:1004.4281 [condmat. mes-hall].
99. Chen D, Zhu H, Liu T. ACS ApplMater Interfaces; 2010. doi:10.1021/am1008437.
100. Das B, Prasad KE, Ramamurty U, Rao CNR. Nanotechnology 2009;20:125705.
101. Liang JJ, Huang Y, Zhang L, Wang Y, Ma YF, Guo TY, et al. Adv Funct Mater 2009;19:2297e302.
102. Zhao X, Zhang QH, Chen DJ, Lu P. Macromolecules 2010;43:2357e63.
103. Yang X, Li L, Shang S, Tao X. Polymer 2010;51:3431e5.
104. Xu YX, Hong WJ, Bai H, Li C, Shi GQ. Carbon 2009;47:3538e43.
105. Jiang L, Shen XP, Wu JL, Shen KC. J Appl Polym Sci 2010;118:275e9.
106. Satti A, Larpent P, Gun'ko Y. Carbon 2010;48:3376e81.
107. Putz KW, Compton OC, Palmeri MJ, Nguyen SBT, Brinson LC. Adv Funct Mater 2010;20:3322e9.
108. Wei T, Luo GL, Fan ZJ, Zheng C, Yan J, Yao CZ, et al. Carbon 2009;47:2296e9.
109. Lee HB, Raghu AV, Yoon KS, Jeong HM. J Macromol Sci Part B Phys 2010;49:802e9.
110. Bryning MB, Milkie DE, Islam MF, Kikkawa JM, Yodh AG. Appl Phys Lett 2005; 87:161909.
111. Paul DR, Robeson LM. Polymer 2008;49:3187e204.
112. Sinha Ray S, Okamoto M. Prog Polym Sci 2003;28:1539e641.

113. Zhang HB, Zheng WG, Yan Q, Yang Y, Wang JW, Lu ZH, et al. Polymer 2010; 51:1191e6.
114. Kim IH, Jeong YG. J Polym Sci Part B Polym Phys 2010;48:850e8.
115. Jiang X, Drzal LT. Polym Compos; 2009:1091e8.
116. Kim S, Do I, Drzal LT. Polym Compos 2010;31:755e61.
117. Kalaitzidou K, Fukushima H, Drzal LT. Compos Part A Appl Sci Manuf 2007; 38:1675e82.
118. Steurer P, Wissert R, Thomann R, Mulhaupt R. Macromol Rapid Commun 2009;30:316e27.
119. Kalaitzidou K, Fukushima H, Drzal LT. Compos Sci Technol 2007; 67:2045e51.
120. Jeong H-K, Lee YP, Jin MH, Kim ES, Bae JJ, Lee YH. Chem Phys Lett 2009;470:255e8.
121. Fim FC, Guterres JM, Basso NRS, Galland GB. J Polym Sci Part A Polym Chem 2010;48:692e8.
122. Jang JY, Kim MS, Jeong HM, Shin CM. Compos Sci Technol 2009;69:186e91.
123. Gu Z, Zhang L, Li C. J Macromol Sci Part B Phys 2009;48:1093e102.
124. Gu ZM, Li CZ, Wang GC, Zhang L, Li XH, Wang WD, et al. J Polym Sci Part B Polym Phys 2010;48:1329e35.
125. Liu P, Gong K, Xiao P, Xiao M. J Mater Chem 2000;10:933e5.
126. Lee SH, Dreyer DR, An JH, Velamakanni A, Piner RD, Park S, et al. Macromol Rapid Commun 2010;31:281e8.
127. Yang YF, Wang J, Zhang J, Liu JC, Yang XL, Zhao HY. Langmuir 2009;25:11808e14.
128. Layek RK, Samanta S, Chatterjee DP, Nandi AK. Polymer 2010;51:5846e56.
129. Goncalves G, Marques PAAP, Barros-Timmons A, Bdkin I, Singh MK, Emami N, et al. J Mater Chem 2010;20:9927e34.
130. Fang M, Wang KG, Lu HB, Yang YL, Nutt S. J Mater Chem 2010;20:1982e92.
131. Zhang B, Chen Y, Zhuang XD, Liu G, Yu B, Kang ET, et al. J Polym Sci Part A Polym Chem 2010;48:2642e9.
132. Li GL, Liu G, Li M, Wan D, Neoh KG, Kang ET. J Phys Chem C 2010;114:12742e8.
133. Zhuang XD, Chen Y, Liu G, Li PP, Zhu CX, Kang ET, et al. Adv Mater 2010;22:1731e5.
134. Park S, Dikin DA, Nguyen ST, Ruoff RS. J Phys Chem C 2009;113:15801e4.
135. Sun ST, Cao YW, Feng JC, Wu PY. J Mater Chem 2010;20:5605e7.
136. Veca LM, Lu FS, Meziani MJ, Cao L, Zhang PY, Qi G, et al. Chem Commun; 2009:2565e7.
137. Coleman J, Khan U, Gun'ko Y. Adv Mater 2006;18:689e706.
138. Akcora P, Kumar SK, Moll J, Lewis S, Schadler LS, Li Y, et al. Macromolecules 2010;43:1003e10.
139. Yang H, Shan C, Li F, Zhang Q, Han D, Niu L. J Mater Chem 2009;19:8856e60.
140. Lee YR, Raghu AV, Jeong HM, Kim BK. Macromol Chem Phys 2009;

210:1247e54.

141. Xu Z, Gao C. Macromolecules 2010;43:6716e23.
142. Huang YJ, Qin YW, Zhou Y, Niu H, Yu ZZ, Dong JY. Chem Mater 2010; 22:4096e102.
143. Liu JQ, Yang WR, Tao L, Li D, Boyer C, Davis TP. J Polym Sci Part A Polym Chem 2010;48:425e33.
144. Liu JQ, Tao L, Yang WR, Li D, Boyer C, Wuhrer R, et al. Langmuir 2010; 26:10068e75.
145. Hu HT, Wang XB, Wang JC, Wan L, Liu FM, Zheng H, et al. Chem Phys Lett 2010;484:247e53.
146. Zheming G, Ling Z, Chunzhong L. J Macromol Sci Part B Phys 2009; 48:226e37.
147. Tkalya E, Ghislandi M, Alekseev A, Koning C, Loos J. J Mater Chem 2010; 20:3035e9.
148. Kim H, Macosko CW. Macromolecules 2008;41:3317e27.
149. Wakabayashi K, Pierre C, Dikin DA, Ruoff RS, Ramanathan T, Brinson LC, et al. Macromolecules 2008;41:1905e8.
150. Wu JH, Tang QW, Sun H, Lin JM, Ao HY, Huang ML, et al. Langmuir 2008; 24:4800e5.
151. Wang J, Ellsworth MW. ECS Trans 2009;19:241e7.
152. Vickery JL, Patil AJ, Mann S. Adv Mater 2009;21:2180e4.
153. Kim H, Macosko CW. Polymer 2009;50:3797e809.
154. Kai W, Hirota Y, Hua L, Inoue Y. J Appl Polym Sci 2008;107:1395e400.
155. Fu X, Qutubuddin S. Polymer 2001;42:807e13.
156. Schaefer DW, Justice RS. Macromolecules 2007;40:8501e17.
157. Yoonessi M, Gaier JR. ACS Nano; 2010. doi:10.1021/nn1019626.
158. Rafiee MA, Rafiee J, Wang Z, Song HH, Yu ZZ, Koratkar N. ACS Nano 2009; 3:3884e90.
159. Li Q, Li ZJ, Chen MR, Fang Y. Nano Lett 2009;9:2129e32.
160. Hirata M, Gotou T, Horiuchi S, Fujiwara M, Ohba M. Carbon 2004;42:2929e37.
161. Fornes TD, Paul DR. Polymer 2003;44:4993e5013.
162. Ansari S, Kelarakis A, Estevez L, Giannelis EP. Small 2010;6:205e9.
163. Schadler LS, Brinson LC, Sawyer WG. J Miner Met Mater Soc 2007;59:53e60.
164. Xu JZ, Chen T, Yang CL, Li ZM, Mao YM, Zeng BQ, et al. Macromolecules 2010;43:5000e8.
165. Salavagione HJ, Martinez G, Gomez MA. J Mater Chem 2009;19:5027e32.
166. Cerezo FT, Preston CML, Shanks RA. Compos Sci Technol 2007;67:79e91.
167. Cai DY, Song M. Nanotechnology 2009;20:315708.
168. Du N, Zhao CY, Chen Q, Wu G, Lu R. Mater Chem Phys 2010;120:167e71.
169. Peponi L, Tercjak A, Verdejo R, Lopez-Manchado MA, Mondragon I, Kenny JM. J Phys Chem C 2009;113:17973e8.

170. Bao QL, Zhang H, Yang JX, Wang S, Tong DY, Jose R, et al. Adv Funct Mater 2010;20:782e91.
171. Solomon MJ, Almusallam AS, Seefeldt KF, Somwangthanaroj A, Varadan P. Macromolecules 2001;34:1864e72.
172. Wagener R, Reisinger TJG. Polymer 2003;44:7513e8.
173. Zhang Q, Fang F, Zhao X, Li Y, Zhu M, Chen D. J Phys Chem B 2008; 112:12606e11.
174. Vermant J, Ceccia S, Dolgovskij MK, Maffettone PL, Macosko CW. J Rheol 2007;51:429e50.
175. Ganguli S, Roy AK, Anderson DP. Carbon 2008;46:806e17.
176. Ramanathan T, Stankovich S, Dikin DA, Liu H, Shen H, Nguyen ST, et al. J Polym Sci Part B Polym Phys 2007;45:2097e112.
177. Ansari S, Giannelis EP. J Polym Sci Part B Polym Phys 2009;47:888e97.
178. Pramoda KP, Linh NTT, Tang PS, Tjiu WC, Goh SH, He CB. Compos Sci Technol; 2009:578e83.
179. Yu A, Ramesh P, Itkis ME, Bekyarova E, Haddon RC. J Phys Chem C 2007; 111:7565e9.
180. Fang M, Zhang Z, Li J, Zhang H, Lu H, Yang Y. J Mater Chem 2010;20:9635e43.
181. Zhang WL, Park BJ, Choi HJ. Chem Commun 2010;46:5596e8.
182. Bansal A, Yang H, Li C, Cho K, Benicewicz BC, Kumar SK, et al. Nat Mater 2005;4:693e8.
183. Priestley RD, Ellison CJ, Broadbelt LJ, Torkelson JM. Science 2005;309:456.
184. Ellison CJ, Torkelson JM. Nat Mater 2003;2:695e700.
185. Rittigstein P, Priestley RD, Broadbelt LJ, Torkelson JM. Nat Mater 2007; 6:278e82.
186. Qiao R, Catherine Brinson L. Compos Sci Technol 2009;69:491e9.
187. Salavagione HJ, Gomez MA, Martinez G. Macromolecules 2009;42:6331e4.
188. Pramoda KP, Hussain H, Koh HM, Tan HR, He CB. J Polym Sci Part A Polym Chem 2010;48:4262e7.
189. Fan H, Wang L, Zhao K, Li N, Shi Z, Ge Z, et al. Biomacromolecules 2010; 11:2345e51.
190. Chen H, Müller MB, Gilmore KJ, Wallace GG, Li D. Adv Mater 2008; 20:3557e61.
191. Balogun YA, Buchanan RC. Compos Sci Technol; 2010:892e900.
192. Roldughin VI, Vysotskii VV. Prog Org Coat 2000;39:81e100.
193. Toker D, Azulay D, Shimoni N, Balberg I, Millo O. Phys Rev B Condens Matter Mater Phys 2003;68:41403.
194. Bauhofer W, Kovacs JZ. Compos Sci Technol 2009;69:1486e98.
195. Martin CA, Sandler JKW, Shaffer MSP, Schwarz MK, Bauhofer W, Schulte K, et al. Compos Sci Technol 2004;64:2309e16.
196. Sandler JKW, Kirk JE, Kinloch IA, Shaffer MSP, Windle AH. Polymer 2003;

44:5893e9.

197. Kovacs JZ, Velagala BS, Schulte K, Bauhofer W. Compos Sci Technol 2007; 67:922e8.
198. Pang H, Chen T, Zhang G, Zeng B, Li ZM. Mater Lett 2010;64:2226e9.
199. Haggenmueller R, Gommans HH, Rinzler AG, Fischer JE, Winey KI. Chem Phys Lett 2000;330:219e25.
200. Du F, Scogna RC, Zhou W, Brand S, Fischer JE, Winey KI. Macromolecules 2004;37:9048e55.
201. Hernández JJ, García-Gutiérrez MC, Nogales A, Rueda DR, Kwiatkowska M, Szymczyk A, et al. Compos Sci Technol 2009;69:1867e72.
202. Wei T, Song L, Zheng C, Wang K, Yan J, Shao B, et al. Mater Lett 2010; 64:2376e9.
203. Hicks J, Behnam A, Ural A. Appl Phys Lett 2009;95:213103.
204. Li J, Kim JK. Compos Sci Technol 2007;67:2114e20.
205. Eda G, Chhowalla M. Nano Lett 2009;9:814e8.
206. Yi YB, Tawerghi E. Phys Rev E Stat Nonlin Soft Matter Phys 2009;79: 041134.
207. Pang HA, Zhang YC, Chen T, Zeng BQ, Li ZM. Appl Phys Lett 2010;96:251907.
208. Liang JJ, Wang Y, Huang Y, Ma YF, Liu ZF, Cai FM, et al. Carbon 2009; 47:922e5.
209. He F, Lau S, Chan HL, Fan JT. Adv Mater 2009;21:710e5.
210. Lee C, Wei X, Kysar JW, Hone J. Science 2008;321:385e8.
211. Paci JT, Belytschko T, Schatz GC. J Phys Chem C 2007;111:18099e111.
212. Rafiee MA, Rafiee J, Srivastava I, Wang Z, Song HH, Yu ZZ, et al. Small 2010;6:179e83.
213. Vaia RA, Maguire JF. Chem Mater 2007;19:2736e51.
214. Kluppel M, editor. The role of disorder in filler reinforcement of elastomers on various length scales, vol. 164. Springer; 2003. p. 86.
215. Akcora P, Liu H, Kumar SK, Moll J, Li Y, Benicewicz BC, et al. Nat Mater 2009;8:354e9.
216. Pukánszky B, Fekete E, editors. Adhesion and surface modification, vol. 139. Springer; 1999. p. 109e53.
217. Lv C, Xue Q, Xia D, Ma M, Xie J, Chen H. J Phys Chem C 2010;114:6588e94.
218. Wagner HD, Vaia RA. Mater Today 2004;7:38e42.
219. Schadler LS, Giannaris SC, Ajayan PM. Appl Phys Lett 1998;73:3842e4.
220. Gong L, Kinloch IA, Young RJ, Riaz I, Jalil R, Novoselov KS. Adv Mater 2010;22:2694e7.
221. Cai M, Glover AJ, Wallin TJ, Kranbuehl DE, Schniepp HC. AIP Conf Proc 2010; 1255:95e7.
222. Frankland SJV, Caglar A, Brenner DW, Griebel M. J Phys Chem B 2002; 106:3046e8.
223. Barber AH, Cohen SR, Wagner HD. Appl Phys Lett 2003;82:4140e2.

224. Wagner HD, Lourie O, Feldman Y, Tenne R. Appl Phys Lett 1998;72:188e90.
225. Yang XM, Tu YF, Li LA, Shang SM, Tao XM. ACS Appl Mater Interfaces 2010;2:1707e13.
226. Cai DY, Yusoh K, Song M. Nanotechnology 2009;20:085712.
227. Nguyen DA, Lee YR, Raghu AV, Jeong HM, Shin CM, Kim BK. Polym Int 2009;58:412e7.
228. Raghu AV, Lee YR, Jeong HM, Shin CM. Macromol Chem Phys 2008; 209:2487e93.
229. Bansal A, Yang H, Li C, Benicewicz BC, Kumar SK, Schadler LS. J Polym Sci Part B Polym Phys 2006;44:2944e50.
230. Khan U, May P, O'Neill A, Coleman JN. Carbon 2010;48:4035e41.
231. Quan H, Zhang B-Q, Zhao Q, Yuen RKK, Li RKY. Compos Part A Appl Sci Manuf 2009;40:1506e13.
232. NguyenDA, RaghuAV,Choi JT, JeongHM. PolymPolym Compos 2010;18:351e8.
233. Verdejo R, Barroso-Bujans F, Rodriguez-Perez MA, de Saja JA, Lopez-Manchado MA. J Mater Chem 2008;18:2221e6.
234. Rafiee MA, Rafiee J, Yu ZZ, Koratkar N. Appl Phys Lett 2009;95:223103.
235. Yavari F, Rafiee MA, Rafiee J, Yu ZZ, Koratkar N. ACS Appl Mater Interfaces 2010;2:2738e43.
236. Rafiq R, Cai D, Jin J, Song M. Carbon 2010;48:4309e14.
237. Jiang X, Drzal LT. Polym Compos 2010;31:1091e8.
238. Liu H, Brinson LC. Compos Sci Technol 2008;68:1502e12.
239. Kim S, Drzal LT. J Adhes Sci Technol 2009;23:1623e38.
240. Prasad KE, Das B, Maitra U, Ramamurty U, Rao CNR. Proc Natl Acad Sci USA 2009;106:13186e9.
241. Jang JY, Jeong HM, Kim BK. Macromol Res 2009;17:626e8.
242. Ciprari D, Jacob K, Tannenbaum R. Macromolecules 2006;39:6565e73.
243. Putz KW, Palmeri MJ, Cohn RB, Andrews R, Brinson LC. Macromolecules 2008;41:6752e6.
244. Jancar J, Douglas JF, Starr FW, Kumar S, Cassagnau P, Lesser AJ, et al. Polymer; 2010:3321e43.
245. Balandin AA, Ghosh S, Bao W, Calizo I, Teweldebrhan D, Miao F, et al. Nano Lett 2008;8:902e7.
246. Ghosh S, Calizo I, Teweldebrhan D, Pokatilov EP, Nika DL, Balandin AA, et al. Appl Phys Lett 2008;92:151911.
247. Seol JH, Jo I, Moore AL, Lindsay L, Aitken ZH, Pettes MT, et al. Science 2010;328:213e6.
248. Yu A, Ramesh P, Sun X, Bekyarova E, Itkis ME, Haddon RC. Adv Mater 2008;20:4740e4.
249. Lin W, Zhang R, Wong CP. J Electron Mater 2010;39:268e72.

250. Veca LM, Meziani MJ, Wang W, Wang X, Lu F, Zhang P, et al. Adv Mater 2009;21:2088e92.
251. Ghose S, Watson KA, Working DC, Connell JW, Smith Jr JG, Sun YP. Compos Sci Technol 2008;68:1843e53.
252. Zhong H, Lukes JR. Phys Rev B 2006;74:125403.
253. Zhang G, Xia Y, Wang H, Tao Y, Tao G, Tu S, et al. J Compos Mater 2010; 44:963e70.
254. Shenogin S, Bodapati A, Xue L, Ozisik R, Keblinski P. Appl Phys Lett 2004; 85:2229e31.
255. Wang SR, Tambraparni M, Qiu JJ, Tipton J, Dean D. Macromolecules 2009; 42:5251e5.
256. Debelak B, Lafdi K. Carbon 2007;45:1727e34.
257. Hung MT, Choi O, Ju YS, Hahn HT. Appl Phys Lett 2006;89:023117.
258. Leszczynska A, Njuguna J, Pielichowski K, Banerjee JR. Thermochim Acta 2007;453:75e96.
259. Kaczmarek H, Podgórski A. Polym Degrad Stab 2007;92:939e46.
260. Mounet N, Marzari N. Phys Rev B Condens Matter Mater Phys 2005;71:205214.
261. Fasolino A, Los JH, Katsnelson MI. Nat Mater 2007;6:858e61.
262. Compton OC, Kim S, Pierre C, Torkelson JM, Nguyen SBT. Adv Mater 2010;22:4759e63.
263. Spitsina NG, Lobach AS, Kaplunov MG. High Energy Chem 2009;43:552e6.
264. Tien CP, Teng HS. J Power Sources 2010;195:2414e8.
265. Liu Z, He D,WangY,WuH,WangJ. Sol Energy Mater Sol Cells; 2010:1196e200.
266. Tung VC, Chen LM, Allen MJ, Wassei JK, Nelson K, Kaner RB, et al. Nano Lett 2009;9:1949e55.
267. Valentini L, Cardinali M, Bon SB, Bagnis D, Verdejo R, Lopez-Manchado MA, et al. J Mater Chem 2010;20:995e1000.
268. Geng X, Niu L, Xing Z, Song R, Liu G, Sun M, et al. Adv Mater 2010;22:638e42.
269. Wu Q, Xu YX, Yao ZY, Liu AR, Shi GQ. ACS Nano 2010;4:1963e70.
270. Wang D-W, Li F, Zhao J, Ren W, Chen Z-G, Tan J, et al. ACS Nano 2009; 3:1745e52.
271. Wang H, Hao Q, Yang X, Lu L, Wang X. Electrochem Commun 2009; 11:1158e61.
272. Yan J, Wei T, Shao B, Fan Z, Qian W, Zhang M, et al. Carbon 2010;48:487e93.
273. Murugan AV, Muraliganth T, Manthiram A. Chem Mater 2009;21:5004e6.
274. Bai H, Xu Y, Zhao L, Li C, Shi G. Chem Commun 2009;45:1667e9.
275. Stejskal J, Gilbert J. Pure Appl Chem 2002;74:857e67.
276. Higashkia S, Kimura K, Matsuo Y, Sugie Y. Carbon 1999;37:351e8.
277. Wang H, Hao Q, Yang X, Lu L, Wang X. ACS Appl Mater Interfaces 2010; 2:821e8.

278. Zhou X, Wu T, Hu B, Yang G, Han B. Chem Commun 2010;46:3663e5.
279. Kulkarni DD, Choi I, Singamaneni SS, Tsukruk VV. ACS Nano 2010;4:4667e76.
280. Williams KA, Boydston AJ, Bielawski CW. J R Soc Interface 2007;4:359e62.
281. Williams KA, Dreyer DR, Bielawski CW. MRS Bull 2008;33:759e65.
282. Xiao X, Xie T, Cheng YT. J Mater Chem 2010;20:3508e14.
283. Kim F, Luo J, Cruz-Silva R, Cote LJ, Sohn K, Huang J. Adv Funct Mater; 2010. doi:10.1002/adfm.201000736.
284. Rothon RN. In: Jancar J, editor. Mineral fillers in thermoplastics: filler manufacture and characterisation, vol. 139. Springer; 1999. p. 67e108.
285. Dasari A, Yu ZZ, Mai YW, Cai GP, Song HH. Polymer 2009;50:1577e87.
286. Wenrong Y, Kyle RR, Simon PR, Pall T, Gooding JJ, Filip B. Angew Chem Int Ed 2010;49:2114e38.
287. Lu J, Do I, Drzal LT, Worden RM, Lee I. ACS Nano 2008;2:1825e32.
288. Lu J, Drzal LT, Worden RM, Lee I. Chem Mater 2007;19:6240e6.
289. Zhou KF, Zhu YH, Yang XL, Luo J, Li CZ, Luan SR. Electrochim Acta 2010; 55:3055e60.
290. Patil AJ, Vickery JL, Scott TB, Mann S. Adv Mater 2009;21:3159e64.
291. Bai H, Li C, Wang X, Shi G. Chem Commun 2010;46:2376e8.
292. Park S, Mohanty N, Suk JW, Nagaraja A, An J, Piner RD, et al. Adv Mater 2010; 22:1736e40.
293. Xie XL, Mai YW, Zhou XP. Mater Sci Eng R Rep 2005;49:89e112.
294. Hussain F, Hojjati M,Okamoto M, Gorga RE. JCompos Mater 2006;40:1511e75.
295. Marcano DC, Kosynkin DV, Berlin JM, Sinitskii A, Sun Z, Slesarev A, et al. ACS Nano 2010;4:4806e14.
296. Segal M. Nat Nanotechnol 2009;4:611e3.
297. UniversityMS http://research.msu.edu/techtransfer/1-mil-support-msu-spinoffxg-sciences. [accessed 30.07.10].
298. Wu X, Qi S, He J, Duan G. J Mater Sci 2010;45:483e9.
299. Vadukumpully S, Paul J, Mahanta N, Valiyaveettil S. Carbon 2011;49:198e205..

Chapter 4

ULTRASONIC SYNTHETIC TECHNIQUE TO MANUFACTURE A PHEMA NANOPOLYMERIC-BASED VACCINE AGAINST THE H6N2 AVIAN INFLUENZA VIRUS: A PRELIMINARY INVESTIGATION

Gérrard Eddy Jai Poinern[1], Xuan Thi Le[1], Songhua Shan[2], Trevor Ellis[3], Stan Fenwick[3], John Edwards[3], and Derek Fawcett[1]

[1]Murdoch Applied Nanotechnology Research Group, Murdoch University, Murdoch, WA, Australia;

[2]Australian Animal Health Laboratories, CSIRO, VIC, Australia;

[3]Veterinary School, School of Veterinary and Biomedical Sciences, Murdoch University, Murdoch, WA, Australia

ABSTRACT

This preliminary study investigated the use of poly (2-hydroxyethyl methacrylate) (pHEMA) nanoparticles for the delivery of the deoxyribonucleic acid (DNA) vaccine pCAGHAk, which expresses the full length hemagglutinin (HA) gene of the avian influenza A/ Eurasian coot/Western Australian/2727/1979 (H6N2) virus with a

Kozak sequence which is in the form of a pCAGGS vector. The loaded and unloaded nanoparticles were characterized using fieldemission scanning electron microscopy. Further characterizations of the nanoparticles were made using atomic force microscopy and dynamic light scattering, which was used to investigate particle size distributions. This preliminary study suggests that using 100 μg of pHEMA nanoparticles as a nanocarrier/adjuvant produced a reduction in virus shedding and improved the immune response to the DNA vaccine pCAG-HAk.

INTRODUCTION

Nanotechnology has been able to deliver a wide range of new and novel materials. These new nanomaterials are increasingly becoming the subject of many investigations in several fields, particularly those of engineering, biotechnology, and biomedical sciences.1,2 Nanomaterials can be made from a wide range of solid materials such as metals, ceramics, polymers, organic materials, and composites. They can come in a wide range of morphologies; namely, spheres, rods, tubes, and plates. The use of nanoparticles of biodegradable polymers is being extensively studied since they provide an attractive alternative for a number of nanomedical applications by providing a delivery platform for the sustained, controlled, and targeted release of drugs and immunogens. These immunogens, therapeutic drug agents, or in some cases imaging agents can be loaded into a biodegradable polymer matrix. Once the polymer is administered, the nanoparticles of the matrix slowly begin to degrade and release the drug or immunogen agents. In addition, the biodegradable nanoparticles can be administered through several different delivery routes, such as oral, nasal, ocular, transdermal, and intravenous routes.3 Ideally these nanoparticles should be inert, biocompatible, and biodegradable. They also need to be stable in vivo, easily attached to immunogens, effectively delivered, and have little or no side effects[4–6]

The 1918 Spanish flu is considered to be the deadliest disaster in human history. The flu killed more than 50 million people worldwide and was related to an avian flu virus H1N1.7 In the late 1990s, the re-emergence of avian influenza demonstrated that this type of virus is persistent and can reach endemic levels in many south-east Asian countries if not effectively managed. Influenza still remains an important and threatening disease to both humans and animals. In contrast to measles, smallpox, and poliomyelitis, influenza is caused by viruses that undergo a continuous antigenic modification within their natural host. The natural reservoir for these viruses is aquatic, migratory birds, which usually flock in large numbers and travel great distances between countries. When these birds associate with the local terrestrial poultry they can occasionally transmit transitory infections. The diversity of diseases that can be transmitted range from mild respiratory illnesses to fatal systematic diseases. The development of new antiviral drugs and vaccines based on nanoparticles has the potential to provide an effective method in dealing with any possible future outbreaks of the influenza virus strains.

Polymer based nanoparticles have been found to improve the therapeutic efficacy and reduce the potential side effects of many therapeutic drug agents. The major challenge facing nanomedicine today in using these polymeric delivery systems is to engineer and manufacture a biodegradable nanoparticle matrix with the desired physiochemical and pharmaceutical properties. If these optimum properties are achieved, then delivering the payload should permit the controlled release of medication concentration within the effective therapeutic window and dosage.[3] The polymeric nanoparticle matrix used in any particular application is an important factor because it can influence parameters such as protein loading, stability, biodegradability, and bioavailability. [8] Thus there are a number of commercially available biodegradable polymers currently used in polymeric nanoparticle matrix formulations. For example, the most widely researched Food and Drug Administration (FDA)-approved biodegradable polymers in the literature are poly (lactide) (PLA), poly (D,L-lactide-co-glycolide) (PLGA) (a copolymer of PLA and poly (glycolide) [PGA]), and poly (ε-caprolactone) (PCL).[9]

The polymers PLA, PLGA, PGA, and PCL were all originally synthesized in the 1950s for nondrug delivery functions such as surgical sutures, textile grafts, and implants. Since this time, these polymers have been also investigated for a variety of drug agent delivery platforms in a number of therapeutic applications. Unfortunately, there are a number of disadvantages in using these polymers; for example, their strong mechanical strength and slow degradation rates, which can lead to a slow drug release that does not provide the desired concentration. In addition, the bioactivity of proteins and peptides encapsulated in the polymer matrix can deteriorate since the polymers' hydrophobic nature can produce an acidic microenvironment. This microenvironment results from water being unable to enter the matrix and the accumulation of acidic breakdown products (lactic and glycolic acid end groups). There are also issues with the hydrophobic nature of the polymeric nanoparticles interacting with hydrophilic molecular probes used for targeting, which can lead to complications in the drug preparation technology.8 It is due to these disadvantages that many researchers look for other novel biodegradable polymers and copolymer delivery platform systems for immunogens and therapeutic drugs.

Many synthetic methods have been used to manufacture a variety of nanopolymeric particles with various sizes and morphologies. The latter parameters having a dominant bearing on the final properties of the nanomaterial synthesized. Some of the attractive features of using a synthetic sonochemical approach are: less complications, reduced processing time, generally more efficient, and economical.[10-13] The sonochemical technique is based on the acoustic cavitation phenomenon, which produces the continuous formation, growth, and final implosive collapse of bubbles in the solution being sonicated. This creates numerous hot-spots in the solution, which provide sufficient energy for the formation and growth of nanoparticles. This synthetic process can be extended to polymers and composite materials.[14,15]

In this article, the development of biocompatible and biodegradable nanosized poly (2-hydroxyethyl methacrylate) (pHEMA) particles that are used as deoxyribonucleic acid (DNA) vaccine carriers is described. DNA vaccination is an effective procedure for inducing protective immunity against a number of infectious and noninfectious diseases in a variety of animals.[16] However, several factors can influence their performance; for example, the delivery technique will dictate the DNA dosage level that is required to solicit an effective immune response. In addition, the rapid degradation and low cellular uptake of plasmid DNA can also have a dramatic effect on the efficiency of the exposed plasmid DNA vaccines.[17] To remediate these problems, plasmid DNA vaccines have been combined with particles, via adsorption, or encapsulation, or by co-formulation to stabilize the plasmid DNA delivery. Combining plasmid DNA with particles significantly reduces the degradation process and also stabilizes the vaccine. It also has the advantage of providing particular materials that are effectively taken up by the antigen presenting cells, thus providing an adjuvant (synergistic) effect.

Historically, adjuvants have been successfully used in the development of vaccines.18 Conventional chemical adjuvants such as bupivacaine19 or Marcaine® (Hospira, Inc, Lake Forest, IL), ubenimex,[20] monophosphoryl lipid A,21 QS-[21] saponin,[22] and levamisole23 have been investigated successfully. These adjuvants were able to facilitate a positive immune response to the DNA vaccines being tested.

The discovery in 1995 of injecting solid inert beads with DNA-encoded antigens resulted in the priming of CD8T cells for a direct immune response.[24] Since then, inert nanoparticles have also been able to induce strong immune responses to protein and peptide antigens in mice[,25,26] sheep,[18] pigs,[27] and cattle.[27] In addition, both metals and inorganic nanomaterials have been used in similar biomedical applications. For example, a metal such as gold which is nontoxic, inert, and stable within the body environment has been used as a contrast agent in cancer diagnosis and photodermal cancer therapy. It has also been effectively used as a delivery platform for oligonucleotide, insulin, and genes.[28] Metal oxide NPs such as magnetite (Fe_3O_4) have been used as magnetically targeted drug delivery platforms due to their biodegradability and biocompatibility. Once these NPs are introduced into the blood stream, the particles flow to the specific location of interest in the body where a strong magnetic field can be used to pull them out of suspension and deliver the pharmaceutical payload.[29] Furthermore, an inorganic material such as mesoporous silica, with its controllable structural properties and biocompatibility, has been effectively used to deliver calcein[.30]

In the case of DNA vaccines, cationic nanoparticles have been formulated with plasmid DNA encoding of a reporter gene. This vaccine was able to enhance the in- vitro cell transfection efficiency and achieve a substantial cellular immune response (16–200 times greater than the normal plasmid DNA by itself) in mice cells via a number of delivery routes.4,31–33 Furthermore, it has also been shown that when both the cholera toxin and lipid A were administered with a nanoparticle- based plasmid DNA, there was an overall synergistic effect which enhanced the immune response of the cellular tissues.5,33 Thus, nanoparticles can also be seen as a novel class of adjuvants, with the potential to be effective delivery platforms for proteins and plasmid DNA immunogens, which can successfully induce a positive immune response. In addition, the nanoparticle delivery platform distributes its payload without the usual side effects associated with local tissue damage caused by conventional chemical adjuvants.

In the past, polymeric particles, both synthetic and natural, have been investigated as potential carriers for the delivery of plasmid DNA to provide cellular immunity. These particles have been found to provide suitable accommodation to plasmids of varying sizes and to provide protection to the plasmid DNA payload from the in-vivo effects of extracellular degradation. In addition, these particles have also provided an enhanced response from the immune system. The first polymeric particle delivery system used for delivering DNA used microsized particles of poly (lactideco-glycolide).34 Recently, polymeric nanoparticles have been investigated for possible plasmid DNA vaccine carriers. The most attractive features of using nanosized polymeric delivery systems are: the nanostructure provides an effective scaffold which is capable of providing a controlled release of DNA, polymeric nanoparticles can easily be manufactured, and they are biocompatible and biodegradable.35,36 Recently, a polymeric microparticle study of formulated plasmid DNA encoding of the nucleoprotein gene A/PR/8/34 of the (H1N1) virus revealed an enhanced immune response in mice37 and a biodegradable pHEMA has also been used in a similar drug delivery system.38,39 However, to date there has been no reports of a nanoparticle pHEMAbased avian influenza DNA vaccine for the H6N2 virus. The present study investigated the immunologic effect of a novel polymeric nanomaterial, pHEMA, as a potential plasmid DNA nanocarrier for a vaccine against the wild bird (H6N2) avian influenza virus. A major advantage of using pHEMA as the vaccine carrier is that this particular nontoxic copolymer has FDA approval for use in contact lenses, implant coatings, and prostheses.

MATERIAL AND METHODS

Chemicals

All chemicals were purchased from Sigma-Aldrich (Castle Hill, NSW, Australia) and used without further purification. Milli-Q® water (18.3 MΩ cm−1) was used throughout all synthesis procedures involving aqueous solutions. The surfactant used in the preparation of the pHEMA nanoparticles was poly (vinyl alcohol) (PVA) and was prepared by dissolving 1 g of PVA in a 100 mL solution of Milli-Q® water. Throughout the preparation of the nanoparticles, a 1% w/v of PVA was used.

Formulation and optimization of solvent

The optimization of the solvent used for mixing pHEMA and DNA, and the calculation of DNA binding, was determined from a comparative study. The study looked at dissolving 1 g of pHEMA in various solutions where the percentage w/v of an alcohol (ethanol) in a Milli-Q water solution was adjusted. In the first case, a 100 mL solution consisted of a mixture composed of 50 mL of Milli-Q water and 50 mL of ethanol; the third solution consisted of a mixture composed of 25 mL of Milli-Q water and 75 mL of ethanol; and the final solution consisted of 100 mL of ethanol. All polymer solvent solutions prepared contained 1% w/v of pHEMA. During the preparation of all loaded and unloaded nanoparticle preparations, 2 mL of pHEMA (1% w/v) was used.

Preparation of pHEMA nanoparticles

The preparation of the unloaded pHEMA nanoparticles started by adding a 2 mL solution of pHEMA (1% w/v) to a 10 mL glass tube supported in an ice bath. The pHEMA was then exposed to ultrasonic irradiation for 30 seconds before 1 mL of PVA (1% w/v) was added to the glass tube dropwise and then sonicated for a further 10 minutes. The ultrasonic processor used throughout these procedures was a UP50H (50 W, 30 kHz, MS7 Sonotrode (7 mm diameter, 80 mm length)) supplied by Hielscher Ultrasound Technology (see Figure 1A).

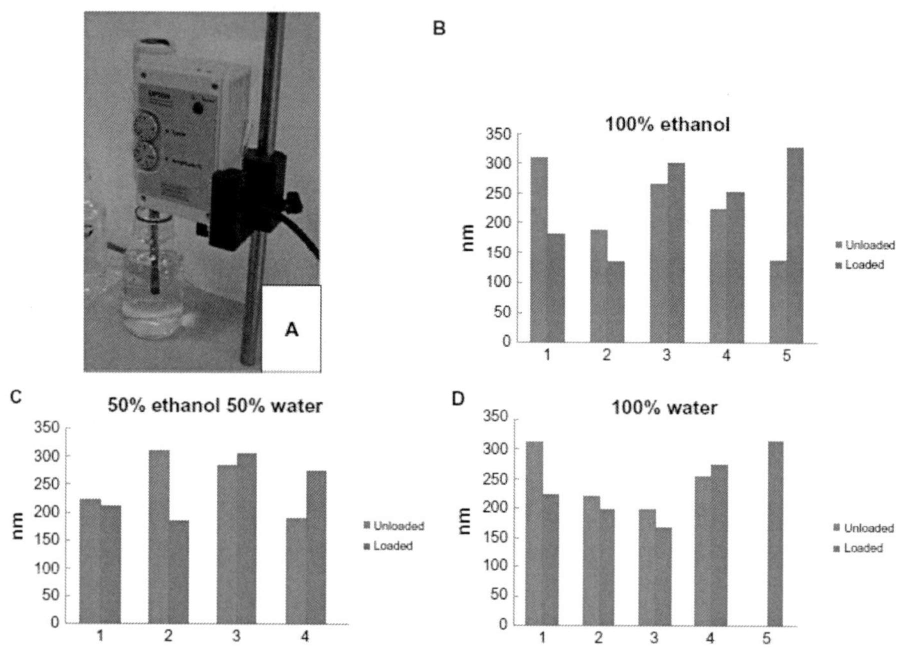

Figure 1. (A) Ultrasonic processor used in the synthesis of nanoparticles. (B–D) Dynamic light scattering data of unloaded and DNA-loaded pHEMA nanoparticles in three solvent mixtures. Abbreviations: DNA, deoxyribonucleic acid; pHEMA, poly (2-hydroxyethyl methacrylate).

Preparation of plasmid DNA vaccine with pHEMA

The formulation of the pHEMA nanoparticles and the DNA composite begins with a 4 mL solution of pHEMA (1% w/v in 100% ethanol blend) being added to a 10 mL glass tube supported in an ice bath. The solution was then sonicated for 30 seconds before a 2 mL solution (1% w/v) of PVA was added dropwise to the glass tube, which was then sonicated for a further 2 minutes. This was followed by the dropwise addition of 400 μL of plasmid DNA (9.3 μg/μL) to the glass tube and then sonicated for a further 10 minutes. At the end of this time, the solution was filtered three times through a 0.2

μm membrane to remove the surfactant. The filtered solution was then centrifuged at 15,000× g for 20 minutes at room temperature. The resultant pellet was then dissolved in phosphate buffered saline (PBS), and the amount of plasmid DNA coated nanoparticles was calculated by subtracting the amount of DNA in the supernatant from the total DNA added. The DNA concentration in the supernatant was measured by a nanodrop ND-1000 Spectrophotometer (Thermo Scientific, Waltham, MA).

Characterization of pHEMA nanoparticles

The structural and morphological features of the dispersed pHEMA nanoparticles were investigated using field-emission scanning electron microscopy (FESEM). Samples were dropped onto a conventional SEM stub and the latter coated with Au. All FESEM scans were taken using a high resolution Zeiss 1555 Variable Pressure Field Emission at 3 kV with a 30 μm aperture under 1 × 10−10 Torr pressure. Atomic force microscopy (AFM) imaging of the pHEMA nanoparticles was carried out by first dropping a few drops of an ethanol solution containing the nanoparticles onto a freshly cleaved mica substrate and then allowing the solution to evaporate before imaging the dry substrate using a Pico-Plus AFM operating in Tapping Mode (molecular imaging). The probes used during the scanning mode were silicon tips with a spring constant of 42 N/m and a resonant frequency of 300 kHz.

The dynamic light scattering (DLS) technique was used to investigate the pHEMA nanoparticle sizes. The loaded and unloaded pHEMA nanoparticles were dispersed in ethanol prior to being investigated by the DLS. The detector used was a Malvern Zetasizer 3000 HAS (Malvern Instruments, Ltd, Worcestershire, UK) (633 nm) operated at 25°C.

Antibody response in animal (chicken) model

For Australian biosecurity reasons, a low pathogenic avian influenza virus, A/Eurasian coot/Western Australia/2727/1979 (H6N2), isolated from a healthy Eurasian coot (Fulica atra) in Australia, was selected to perform this DNA vaccine adjuvant study. The procedure of combining the pCAG-HAk plasmid DNA that expresses the complete hemagglutinin (HA) gene of the avian influenza virus (H6N2), together with a Kozak sequence, in a pCAGGS vector used in the DNA vaccine has been previously described by Shan et al.[40]

Vaccination regime

All bird experiments were carried out with the approval of Murdoch University's animal ethics committee, and all experiments were conducted in accordance with the Australian National Health and Medical Research Council's (NHMRC) code of practice for the care and use of animals for scientific purposes. The birds selected for this study were 3-week-old Hy-Line chickens that were free from the avian influenza. The chickens were accommodated in free-range pens with access to feed and water, and were maintained at the Animal Resource Centre, Murdoch University, Perth, Western Australia.

The experimental immunization protocol used in the bird vaccine study is presented in Table 1 and contains information regarding bird numbers and vaccine dosage. During the protocol, each bird received two intramuscular injections of 0.2 mL at 3-week intervals. The injection procedure involved a 0.1 mL dose being injected in each leg. Over this period, the six control birds received a 200 µL dose of PBS without adjuvant, five birds received a 100 µL dose of pCAG-HAk without adjuvant, and the remaining sets of birds received doses of 10 µL, 100 µL, and 200 µL of pCAG-HAk with pHEMA. Sera samples were collected weekly to detect and monitor the H6 specific antibody of hemagglutination-inhibition (HI).40,41 Three weeks after the booster vaccination, each bird received a 0.5 mL dose of the wild bird H6N2 avian influenza virus (106.5 EID50/0.1 mL) via three delivery routes. In the first route, the bird received 0.1 mL of the H6N2 virus by nasal instillation, the second 0.1 mL was introduced via eye drops, and the final 0.3 mL dose was delivered through an oral route. Following the virus challenge, a daily observation of all birds was undertaken and either oropharyngeal or cloacal swabs were collected every second day over a 7-day period. The virus isolation procedure was performed in accordance with the manual of diagnostic tests and vaccines for terrestrial animals.[41]

RESULTS AND DISCUSSION

Optimization of the solvent revealed that the pHEMA nanoparticles prepared with 100% ethanol produced the highest DNA binding rate. The total plasmid DNA used in each solvent experiment was 581.3 ng/μL, and the maximum DNA binding rate was found in 100% ethanol solution, with only 28.3 ng/μL of DNA remaining in the supernatant (see Table 2). Thereafter, 100% ethanol was used for the preparation of the pHEMA adjuvant vaccine for the rest of the study.

In the 100% water case, the unloaded particles of pHEMA ranged in size from 168 to 314 nm, with the unloaded particles being larger in most cases. In the 50% ethanol and 50% water case, the pHEMA particles ranged in size from 185 to 311 nm, Table 2 The effect of solvent composition on plasmid DNA binding to pHEMA with the unloaded particles being larger in half of the trials.

Table 1. Immunization protocol in chicken vaccine study

Group	Vaccine	Dose, μg	Adjuvant	Number of chickens
1	PBS	100	None	6
2	pCAG-HAk	100	None	5
3	pCAG-HAk	10	pHEMA	5
4	pCAG-HAk	100	pHEMA	3
5	pCAG-HAk	200	pHEMA	4

Table 2. The effect of solvent composition on plasmid DNA binding to pHEMA

Solvent	100% Milli-Q® water	50% Milli-Q® water 50% ethanol	100% ethanol
DNA concentration in supernatant, ng/μL	505.9	208.8	28.30
Bonded DNA, ng/μL	75.40	372.5	553.0
Bonded DNA, %	13.00	64.10	95.10

Abbreviations: DNA, deoxyribonucleic acid; pHEMA, poly (2-hydroxyethyl methacrylate).

While in the 100% ethanol case, the pHEMA particles ranged from 136 to 328 nm, with the unloaded particles being smaller in more than half of the trials (see Figure 1). Figure 2 presents FESEM images of the unloaded and DNA-loaded pHEMA nanoparticles; the images reveal that the nanoparticles range in size from 120 to 330 nm and have spherical morphology. This is confirmed by the AFM profile images of the unloaded and loaded pHEMA nanoparticles presented in Figure 3. The DLS, FESEM, and AFM analysis all confirm that the nanoparticles range in size from 120 to 330 nm; the morphology is spherical in the unloaded system and appears to be unchanged by the incorporation of the DNA.

Following DNA loading procedures, the efficiency of the nanovaccine was tested in a bird (Hy-Line chicken) model. Three weeks post second vaccination, no H6 HI antibody titer was detected in any of the Hy-Line chickens. At the end of 10 days post virus challenge, all birds sero-converted with a range of HI titers, which are presented in Table 3. There was a

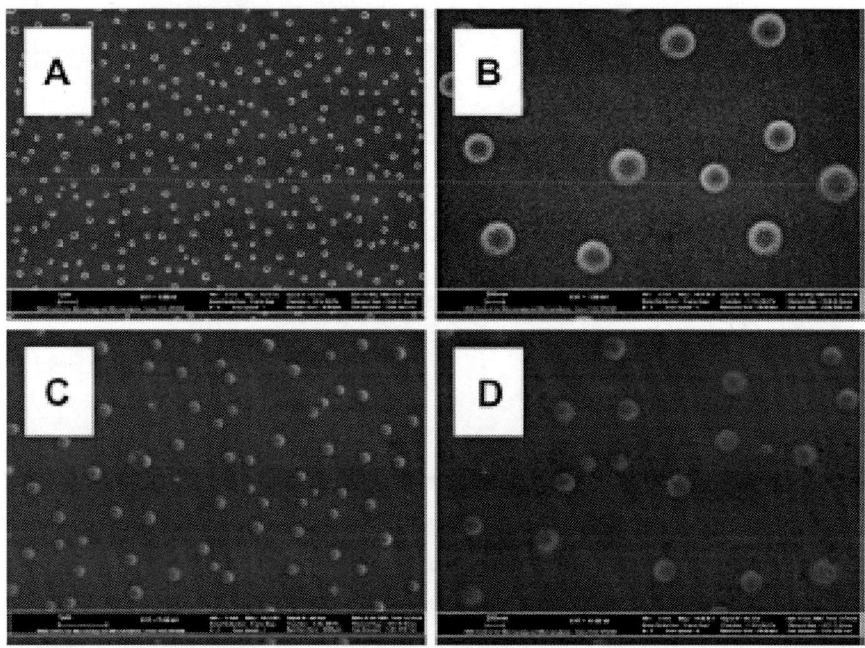

Figure 2 Field-emission scanning electron microscopy images of the unloaded and DNA-loaded pHEMA nanoparticles with the spherical particle morphology. (A) Unloaded pHEMA nanoparticles at low magnification (scale bar 1 μm) and (B) at high magnification (scale bar 200 nm); and (C) DNA-loaded pHEMA nanoparticles at low magnification (scale bar 1 μm) and (D) at high magnification (scale bar 200 nm). Abbreviations: DNA, deoxyribonucleic acid; pHEMA, poly (2-hydroxyethyl methacrylate).

Figure 3. Atomic force microscopy profile images of the unloaded and DNA-loaded pHEMA nanoparticles with spherical particle morphology. (A) Unloaded pHEMA nanoparticles at low magnification and (B) at high magnification; and (C) DNA-loaded nanoparticles of pHEMA at low magnification and (D) at a higher magnification. Abbreviations: DNA, deoxyribonucleic acid; pHEMA, poly (2-hydroxyethyl methacrylate). significant difference (P , 0.05) in the geometric mean titer of the HI antibody prior to and post virus challenge using the paired-sample t-test.

Table 4 presents the level of virus shedding in both the oropharyngeal and cloacal swabs in vaccinated birds following the H6N2 avian influenza virus challenge.

Comparing the pCAG-HAk vaccinated group with the naive control group, we can see that the virus excretion rate in the oropharyngeal swabs of 70.8% in the naive group was reduced to 45% in the pCAG-HAk group following the virus challenge. While cloacal swabs for the naive control group recorded a 12.5% value, the pCAG-HAk vaccinated group was reduced to 0% value for the post virus challenge. Table 4 reveals that there is a significant difference between the pCAG-HAk vaccinated group and the naive control group for both oropharyngeal and cloacal swabs. In comparison with the naive control group, the 100 μg pHEMA adjuvanted plasmid pCAG-HAk group has shown a significant decrease in virus shedding in both oropharyngeal and cloacal swabs. This result suggests that using pHEMA as;

Table 3. Antibody response prior to and post virus challenge

Time	Naive control	pCAG-HAk			
		No pHEMA (adjuvant)	pHEMA, μg		
Prior	0	0	10	100	200
Post[a]	6.0 ± 2.0	5.6 ± 1.7	6.6 ± 2.1	5.3 ± 1.5	5.3 ± 1.5

Note: aValues represent geometric mean titer (log2) ± standard deviation of each bird group. Abbreviation: pHEMA, poly (2-hydroxyethyl methacrylate).

Table 4. Virus shedding level in chickens vaccinated with plasmid pCAG-HAk and pHEMA adjuvants following the virus challenge

Swabs	Naive control	pCAG-HAk			
		No pHEMA (adjuvant)	pHEMA, μg		
			10	100	200
OS	17/24 (70.8)	9/20 (45)	9/20 (45)	3/12 (25)	8/16 (50)
CS	3/24 (12.5)	0/20 (0)	1/20 (5)	0/12 (0)	1/16 (6.3)

Notes: Values are number of positive swabs for virus isolation per total number of swabs tested.

Percentage rate is shown in parentheses. Abbreviations: CS, cloacal; OS, oropharyngeal; pHEMA, poly (2-hydroxyethyl methacrylate). An adjuvant has improved the immune response to the DNA vaccine pCAG-HAk. It is suspected that a major factor that contributed to the improved immune response produced by the pHEMA lies in its hydrophilic nature and its positive surface charge that was conveyed from the ethanol solvent. This positive surface charge significantly improves the absorption of plasmid DNA than the water solvent, see Table 2. The plasmid DNA is believed to be absorbed onto the surface of pHEMA nanoparticles through an electrostatic interaction or by covalent binding. Importantly, this bonding mechanism did not appear to damage the plasmid DNA's immunization performance. In addition, the bonding mechanism appears to be complex because the optimum effect occurred in the bird group that received the 100 µg dose. Both the 10 and 200 µg dose groups were less effective. The mechanisms behind the observed adjuvant effect have not yet been resolved.

Some possible mechanisms that could have affected the performance of the pHEMA nanoparticle delivery platform include: the surface distribution of the plasmid DNA on the nanoparticles, the delivery platform prevented effective DNA degradation, which in turn affected the targeting of the DNA to antigen presenting cells.34,42 Another factor that needs to be investigated further is the effect of the nanoparticle size used in the delivery platform. In this study, the pHEMA nanoparticle size ranged from 120 to 330 nm, and this size range could have influenced the delivery mechanism. Recent studies have shown that the size of the nanoparticles being used as the DNA delivery platforms can have a significant effect on the DNA vaccine efficacy.[5,18,25,43]

The use of particular delivery platforms as a novel method of delivering a payload of proteins and/or plasmid DNA immunogens to induce a positive immune response is a research area that is currently receiving a great deal of interest. The characteristics of both micro- and nanosized particles can have a significant impact on the overall performance of the delivery system. For example, the size, shape, and surface properties such as hydrophobicity and surface charge directly affect the efficiency of the particular delivery system.44 Recently, a variety of inert nanoparticles was investigated and found to be effective delivery vehicles for protein and peptide antigens.4,28–30,39,45,46 However, the application of nanoparticles as delivery platforms with DNA vaccines as payloads is only at the exploratory stage.33,43 In this present preliminary study, the potential application of using a biodegradable, nontoxic copolymer (pHEMA nanoparticles) as a delivery platform to carry pCAG-HAk plasmid DNA has been investigated. In the past, pHEMA has been used in drug delivery systems;26,31 however, to date, it appears that this study is the first to use pHEMA nanoparticles as an adjuvant in DNA vaccination.

CONCLUSION

This preliminary study suggests that using pHEMA nanoparticles as a nanocarrier/adjuvant have improved the immune response to the DNA vaccine pCAG-HAk. A reduction in virus shedding was detected in both oropharyngeal and cloacal swabs for the 100 µg pHEMA adjuvant DNA vaccine. Three pHEMA adjuvant doses (10, 100, and 200 µg) were investigated. The study revealed that there was a dose response effect, with the 100 µg producing the most significant amount of virus shedding. The mechanism behind this adjuvant effect has not been resolved, but the reduction of virus shedding in the oropharynx of chickens challenged with the wild bird H6N2 influenza virus warrants further investigation.

ACKNOWLEDGMENTS

The authors would like to thank Dr Zhong-Tao Jiang for his helpful discussions. Ms Jennifer Millar is also acknowledged for her assistance. Dr Derek Fawcett would like to thank the Bill and Melinda Gates Foundation for his research fellowship.

Disclosure

The authors report no conflict of interest in this work.

REFERENCES

1. Ramakrishna S, Ramalingam M, Sampath Kumar TS, and Soboyejo WO. Biomaterials: a nano approach. 1st ed. Boca Raton, FL: CRC Press; 2010.
2. Greco RS, Prinz FB, Smith RL. Nanoscale Technology in Biological Systems. 1st ed. Boca Raton, FL: CRC Press; 2005.
3. Feng SS. New concept chemotherapy by nanoparticles of biodegradable polymers: where are we now? Nanomedicine. 2006;1(3):297–309.
4. Cui Z, Mumper RJ. Genetic immunization using nanoparticles engineered from micro-emulsion precursors. Pharm Res. 2002;19:939–946.
5. Wang R, Doolan DL, Le TP, Hedstrom RC, Coonan KM. Induction of antigen-specific cytotoxic T lymphocytes; in humans by a malaria DNA vaccine. Science. 1998;282:476–480.
6. Scheerlinck JP, Greenwood DL. Particulate delivery systems for animal vaccines. Methods. 2006;40:118–124.
7. Patterson KD, Pyle GF. The geography and mortality of the 1918 influenza pandemic. Bull Hist Med. Spring 1991;65(1):4–21.
8. Lee M, Iruela Arispe MI, Wu BM, Dunn JCY. Modulation of protein delivery from modular polymer scaffolds. Biomaterials. 2007;28(10): 1862–1870.
9. Z, Feng SS. In vitro investigation on poly (lactide) Tween 80 copolymer nanoparticles by dialysis method for chemotherapy. Biomacromolecules. 2006;7:1139–1146.
10. Suslick KS, Fang MM, Hyeon T. Nanostructured materials generated by high intensity ultrasound: sonochemical synthesis and catalytic studies. J Am Chem Soc. 1996;118:2172–2179.
11. Bang JH, Suslick KS. Applications of ultrasound to the synthesis of nanostructured materials. Adv Mater. 2010;22:1039–1059.
12. Geng J, Liu B, Xu L, Hu FN, Zhu JJ. Facile route to Zn based II–VI semiconductor spheres, hollow spheres and core/shell nanocrystals and their optical properties. Langmuir. 2007;23(20):10286–10293.
13. Poinern GEJ, Brundavanama RK, Mondinosa N, Jiang ZT. Synthesis and characterisation of nanohydroxyapatite using an ultrasound assisted method. Ultrason Sonochem. 2009;16(4):469–474.
14. Zhang K, Park BJ, Fang FF, Choi HJ. Sonochemical preparation of polymer nanocomposites. Molecules. 2009;14:2095–2110.
15. Freitas S, Hielscher G, Merkle HP, Gander B. A fast and simple method for producing biodegradable nanospheres. Eur Cell Mater. 2004; 7(2):28–29.
16. Abdulhaqq SA, Weiner DB. DNA vaccines: developing new strategies to enhance immune responses. Immunol Res. 2008;42:219–232.
17. Wilson KD, de Jong SD, Kazem M, et al. The combination of stabilized plasmid lipid particles and lipid nanoparticle encapsulated CpG containing oligodeoxynucleotides as a systemic genetic vaccine. J Gene Med. 2009;11:14–25.

18. Scheerlinck JP, Gloster S, Gamvrellis A, Mottram PL, Plebanski M. Systemic immune responses in sheep, induced by a novel nano-bead adjuvant. Vaccine. 2006;24:1124–1131.
19. Wang B, Ugen KE, Srikantan V, et al. Gene inoculation generates immune responses against human immunodeficiency virus type 1. Proc Natl Acad Sci U S A. 1993;90:4156–4160.
20. Sasaki S, Fukushima J, Hamajima K, et al. Adjuvant effect of Ubenimex on a DNA vaccine for HIV-1. Clin Exp Immunol. 1998;111:30–35.
21. Sasaki S, Hamajima K, Fukushima J, et al. Comparison of intranasal and intramuscular immunization against human immunodeficiency virus type 1 with a DNA-monophosphoryl lipid A adjuvant vaccine. Infect Immun. 1998;66:823–826.
22. Sasaki S, Sumino K, Hamajima K, et al. Induction of systemic and mucosal immune responses to human immunodeficiency virus type 1 by a DNA vaccine formulated with QS-21 saponin adjuvant via intramuscular and intranasal routes. J Virol. 1998;72:4931–4939.
23. Jin H, Li Y, Ma Z, et al. Effect of chemical adjuvants on DNA vaccination. Vaccine. 2004;22:2925–2935.
24. Falo LD, Kovacsovics-Bankowski M, Thompson K, Rock KL. Targeting antigen into the phagocytic pathway in vivo induces protective tumour immunity. Nat Med. 1995;1:649–653.
25. Fifis T, Gamvrellis A, Crimeen-Irwin B, et al. Size-dependent immunogenicity: therapeutic and protective properties of nano-vaccines against tumours. J Immunol. 2004;173:3148–3154.
26. Fifis T, Mottram P, Bogdanoska V, Hanley J, Plebanski M. Short peptide sequences containing MHC class I and/or class II epitopes linked to nano-beads induce strong immunity and inhibition of growth of antigenspecific tumour challenge in mice. Vaccine. 2004;23:258–266.
27. Aucouturier J, Dupuis L, Ganne V. Adjuvants designed for veterinary and human vaccines. Vaccine. 2001;19:2666–2672.
28. Misha B, Patel BB, Tiwari S. Colloidal nanocarriers: a review on formulation technology, types and applications towards targeted drug delivery. Nanomedicine. 2010;6:9–24.
29. Pankhurst QA, Connolly J, Jones SK, Dobson J. Applications of magnetic nanoparticles in biomedicine. J Phys D Appl Phys. 2003; 16:167–181.
30. Slowing II, Vivero-Escoto JL, Wu C, Lin VS. Mesoporous silica nanoparticles as controlled release drug delivery and gene transfection carriers. Adv Drug Deliv Rev. 2008;60:1278–1288.
31. Cui Z, Mumper RJ. Intranasal administration of plasmid DNAcoated nanoparticles results in enhanced immune responses. J Pharm Pharmacol. 2002;54:1195–1203.
32. Cui Z, Mumper RJ. Topical immunization using nanoengineered genetic vaccines. J Control Release. 2002;81:173–184.

33. Cui Z, Mumper RJ. The effect of co-administration of adjuvants with a nanoparticle-based genetic vaccine delivery system on the resulting immune responses. Eur J Pharm Biopharm. 2003;55:11–18.
34. Singh M, Briones M, Ott G, O'Hagan D. Cationic microparticles: a potent delivery system for DNA vaccines. Proc Natl Acad Sci U S A. 2000;97:811–816.
35. Nguyen DN, Green JJ, Chan JM, Langer R, Anderson DG. Polymeric materials for gene delivery and DNA vaccination. Adv Mater. 2009; 21:847–867.
36. Galindo-Rodriguez SA, Allemann E, Fessi H, Doelker E. Polymeric nanoparticles for oral delivery of drugs and vaccines: a critical evaluation of in vivo studies. Crit Rev Ther Drug Carrier Syst. 2005; 22:419–464.
37. Hartikka J, Geall A, Bozoukova V, et al. Physical characterization and in vivo evaluation of poloxamer-based DNA vaccine formulations. J Gene Med. 2008;10:770–782.
38. Piotrowicz A, Shoichet MS. Nerve guidance channels as drug delivery vehicles. Biomaterials. 2006;27:2018–2027.
39. Rao KP. New concepts in controlled drug delivery. Pure Appl Chem. 1998;70:1283–1287.
40. Shan SH, Jiang YP, Bu ZG, et al. Strategies for improving the efficacy of a H6 subtype avian influenza DNA vaccine in chickens. J Virol Methods. 2011;173:220–226.
41. O.i.d.é. Avian Influenza. In: Manual of Diagnostic Tests and Vaccines for Terrestrial Animals Office international des epizooties. http://www. oie. int/eng/normes/mmanual/2008/pdf/2.03.04.AI.pdf.
42. Khatri K, Goyal AK, Vyas SP. Potential of nanocarriers in genetic immunization. Recent Pat Drug Deliv Formul. 2008;2:68–82. 43.
43. Minigo G, Scholzen A, Tang CK, et al. Poly-L-lysinecoated nanoparticles: a potent delivery system to enhance DNA vaccine efficacy. Vaccine. 2007;25:1316–1327.
44. Xiang SD, Scholzen A, Minigo G, et al. Pathogen recognition and development of particulate vaccines: does size matter? Methods. 2006;40:1–9.
45. Li GP, Liu ZG, Liao B, Zhong NS. Induction of Th1-type immune response by chitosan nanoparticles containing plasmid DNA encoding house dust mite allergen Der p 2 for oral vaccination in mice. Cell Mol Immunol. 2009;6:45–50.
46. Lori F, Calarota SA, Lisziewicz J. Nanochemistry-based immunotherapy for HIV-1. Curr Med Chem. 2007;14:1911–1919..

Chapter 5

PROTECTIVE EFFECT OF CONVENTIONAL ANTIOXIDANT (!-CAROTENE, RESVERATROL AND VITAMIN E) IN CHITOSAN-CONTAINING HYDROGELS AGAINST OXIDATIVE STRESS AND REVERSAL OF DNA DOUBLE STRANDED BREAKS INDUCED BY COMMON DENTAL COMPOSITES: IN-VITRO MODEL

V. Tamara Perchyonoka,*, Shengmiao Zhangb, and Theunis Oberholzerc

[a]Vtpchem Pty Ltd, Glenhuntly, Melbourne, 3163, Australia

[b]Shanghai Key Laboratory of Advanced Polymeric Materials, Key Laboratory for Ultrafine Materials of Ministry of Education, School of Materials Science and Engineering, East China University of Science and Technology, Shanghai 200237, China

[c]School of Dentistry and Oral Health, Griffith University, 4125, QLD, Australia

ABSTRACT

Cytotoxic resin components of common dental bonding agents are known to cause oxidative damage and suppress odontogenic differentiation of dental pulp cells. [1-4] As antioxidants were found to protect cells from cytotoxicity of resin monomers in previous studies,[5-8] we investigated the effects of common antioxidants, such as !-carotene, resveratrol and Vitamin E on anti-differentiation activity of bonding agents without compromising bond strength. Methacrylate monomers used in dentistry have been shown to induce DNA double strand breaks (DSBs), a severe type of DNA damage. The formation of classical products of oxidative DNA damage, were studied using the UV-detection method.

We found that exposure of 2′-deoxyguanosine, phenylalanine and 2′-deoxythymene to a commercially available dental composite restorative material (Esthet-X micromatrix restorative composite material) lead to various degrees of formation of 8-oxo-7,8-dihydro-2′-deoxyguanosine (8oxodG), tyrosine and 5-hydroxy-2′-deoxythymidine (5OHdT) as essential markers of oxidative DNA double stranded damage. The yields of 8oxodG, tyrosine and 5OHdT appear to be negligible in the presence of antioxidant containing chitosan hydrogels. We also extended the methodology to hydrogen atom transfer reactions of 2-bromo naproxen methyl ester and 2-bromoibuprofen methyl ester as important class of radical forming reactions under blue light conditions and found that Gels play an important protective role under identical conditions Gels were characterized with SEM and the images incorporated.

Conclusion: Antioxidant containing chitosan hydrogels may reduce detrimental effects induced by common composite restorative agent in vitro and introducing additional therapeutic health benefits.

INTRODUCTION

Modern dental adhesive systems are used to improve contact between restorative material and the walls of the prepared cavity of the tooth. As these materials come in close and prolonged contact with vital dentin, their influence on pulp tissue is critical. Thus, the biocompatibility of dentin bonding agents is a relevant aspect of the clinical success of these materials [1-3]. Dentin bonding agents alone proved to be cytotoxic [4], and it has been found that the type and quantity of leachable components significantly influence the biological behavior of resin restorations [5-10]. Cytotoxicity of dentin bonding agents has been examined using a variety of cell lines including primary human pulp and pulp-derived cells [4, 5-8, 11-14].

As a consequence of aerobic metabolism, small amounts of reactive oxygen species (ROS) are constantly generated in cells and tissues. Cellular antioxidants like glutathione act in unison to detoxify these reactive molecules, but when the balance between oxidants and antioxidants is disrupted, a condition referred to as oxidative stress exists. If oxidative stress persists, oxidative damage to lipids, proteins and nucleic acids accumulates and eventually results in cell death [4]. Co-polymeric and monomeric form resin-based materials such as HEMA and TEGDMA are a likely cause of cellular damage via the uncontrollable generation of reactive oxygen species. Recently, the possible link between the highly reactive and poorly controlled free radical species generation and cytotoxycity has been established [15-19]. Also genotoxic effects of triethyleneglycol dimethacrylate been demonstrated in vitro as well, indicating DNA reactivity of the compounds [4, 20, 21].

Table 1. Gel formulation prepared in the Study

Gel formulation		Medium	pH
Chitosan-H	Gel-1	3% acetic acid	4.00
Chitosan-H+Resveratrol	Gel-2	3% acetic acid	4.16
Chitosan-H+Vitamin E	Gel-3	3% acetic acid	4.63
Chitosan-H+β-carotene	Gel-4	3% acetic acid	5.12

The human body has several mechanisms to counteract damage by free radicals and other reactive oxygen species. These act on different oxidants as well as in different cellular compartments. The presence of antioxidants is the fundamental line of defence against free radical damage. An antioxidant is a molecule stable enough to donate an electron to a reactive free radical and quench it, thus reducing its capacity to damage. Some such antioxidants, including glutathione, ubiquinol and uric acid, are produced during normal metabolism in the body. Other lighter antioxidants are found in the diet. Although about 4000 antioxidants have been identified, the best known are vitamin E, vitamin C and the carotenoids. Many other non-nutrient food substances, generally phenolic or polyphenolic compounds, display antioxidant properties and, thus, may be important for health.

Although a wide variety of antioxidants in foods contribute to disease prevention, the bulk of research has focused on three antioxidants, which are essential nutrients or precursors of nutrients. These are vitamin E, vitamin C and the carotenoids. Each of these antioxidant nutrients have specific activities and they often work synergistically to enhance the overall antioxidant capability of the body [15-17]. The balance between the production of free radicals and the antioxidant defences in the body has important health implications. Chitosan, which is a biologically safe biopolymer as well as an antioxidant, has been proposed as a bioadhesive polymer and is of continuous interest to us due to its unique properties and flexibility in broad range of oral applications reported by others and us recently [15-18].

The main objective of this study was to evaluate the effect of 3 chitosan-antioxidant hydrogels on the antioxidant defence mechanism (resveratrol, vitamin E and !-carotene on in-vitro model of oxidative damage potentially generated by the model composite. Secondly, we aimed to investigate the chemical nature of the defence on the interface between the composites and anioxidant/chitosan hydrogel layer formation by the use of SEM.

MATERIAL AND METHODS

Materials

!-carotene (Aurora Pharmaceuticals, Australia), resveratrol (Nature's care manufacture PTY LTD, Australia), vitamin E (Now

Food, Australia), 2′-deoxyguanosine (Sigma, USA), phenylalanine (Sigma, USA) and 2′-deoxycytidine (Sigma, USA) were used as received. Chitosan (Aldrich, Australia), glycerol (Sigma, USA), glacial acetic acid (E.Merck, Germany), naproxen and ibuprofen (Sigma, USA) were used as received. The degree of deacetylation of typical commercial chitosan used in this study is 87%. Chitosan with molecular weight 2.5x 103 KD was used in the study. Gelatin in powder form was purchased from Shanghai Chemical Reagent Co., (Shanghai, China) with the numberaverage molecular weight (Mn) of about 8.7 x104. The isoelectric point is 4.0-5.0.

Preparation of chitosan based gels with resveratrol, !- carotene or vitamin E as suitable bio-molecular drug delivery systems and a hybrid layer formation promoter.

Antioxidant containing chitosan gel was prepared by dispersion of corresponding antioxidant 0.2 gm in glycerol (5% w/w) using a mortar and a pestle. Ten milliliters of glacial acetic acid (3% w/w) was then added with continuous mixing and finally chitosan polymer was spread on the surface of the dispersion and mixed well to form the required gel. Antioxidant gel had been prepared with three different 2 w/w concentrations of chitosan gelling agent and the summary of the newly prepared materials is highlighted in Table 1. Glacial acetic acid was added to the preparation to obtain the homogenous clear chitosan gel.

Determination of Gel pH

One gram of the prepared gels was accurately weighed and dispersed in 10 ml of purified water. The pH of the dispersions was measured using pH meter (HANNA instruments, HI8417, Portugal) [12].

Morphology of the Gels

The samples were prepared by freezing in liquid nitrogen for 10 min, and then were freeze-dried for 24 h. The prepared samples were fractured in liquid nitrogen using a razor blade. The fractured samples were attached to metal stubs, and sputter coated with gold under vacuum for SEM. The interior and the surface morphology were observed in scanning electron microscope (SEM, Hitachi S4800, Japan).

H_2O_2/Cupric Ions (Cu^{2+}) Hydroxyl Radical Generation in the In Vitro Conditions. In order to further measure the reducing ability of negative control (35% hydrogen peroxide solution and $CuSO_4$), vitamin C, vitamin E, resveratrol and beta-carotene, the cupric ions' (Cu^{2+}) reducing power capacity was used with slight modification [33]. Briefly, 250 μL of 37.5% hydrogen peroxide solution and $CuSO_4$ and 250 μL CH_3COONH_4 buffer solution (100 mmol/L, pH 7.0) were added to a test vial containing a negative control (35% hydrogen peroxide solution and $CuSO_4$) vitamin E, resveratrol and !-carotene sample as well as chitosan-antioxidant gels of the vitamin E, resveratrol and !-carotene (250 μL). Then, the total volume was adjusted with the buffer to 2 mL and mixed vigorously.

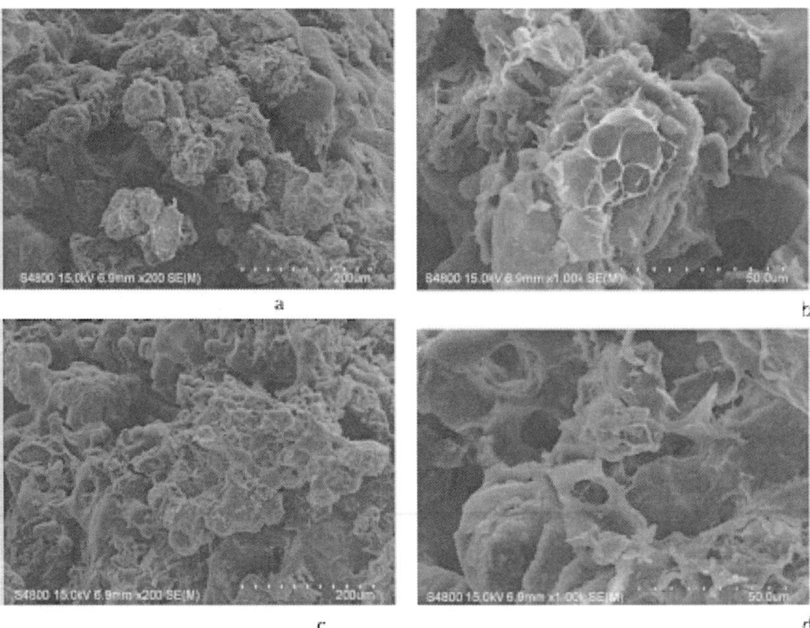

Figure 1. SEM photographs of interior morphology of the selected gels under investigation for (a) Gel-1, (b) Gel-2 (c) Gel-3, (d) Gel-4.

Absorbance against a buffer blank was measured at 568 nm at 20 minutes intervals for the total time of 2 hours. The gels were tested after 1 to 14 weeks in order to establish the longevity and the importance of stability of the gel compostion. Increased absorbance of Cu+ complex in the reaction mixture indicates increased reduction capability. Trolox (water soluble Vitamin E) was used as a positive control. The results of the investigation are summarized in graph 2. Absorbance was measured using POLARstar Omega Multifunction Microplate Reader (BMG LABTECH, Spectral range: 220 - 850 nm). 24 well plates used in the investigations are Corning Incorporated Castar 3524, 24 well cell culture cluster flat bottom with lid, Non-pyrogenic, Polystyrene, sterile plates. (Corning Incorporated Corning, NY, 14831, USA).

General Experimental Procedure for Hydrogen Atom Transfer Reaction Under Blue Light Conditions at Ambient Temperature:

To a solution of H_3PO_2 (20% solution in water, 5 equiv) and Bu4N+Cl- (5 equiv) in water (5 ml) was added radical precursor (0.1 mmol) followed by irradiation with conventional composite curing light for 1h. The solution was stirred at room temperature for 1h. The reaction mixture was then extracted with EtOAc (2x 10 ml) and the organic phase dried with $MgSO_4$. The crude mixture did not require further purification and was analyzed by ^1H, gCOSY NMR spectroscopy, and HPLCMS to confirm formation of the reduced product.

RESULTS AND DISCUSSION

Table 1 represents the summary of the additive-chitosan gels prepared and used in this study. Additive (active antioxidant) content in 0.3 g of different gel formulations from the prepared formulae (presented in Table 1). The prepared gel formulations have uniform distribution of drug content, homogenous texture and yellow color. The pH of the formulations ranged from to 4-5.12.

The SEM images were obtained to characterize the microstructure of the freeze-dried additive composite gels (gel1-4) and are presented in Fig. (1). The gels displayed a homogeneously porous structure. It was thought that the micro- porous structure of the gels could lead to high internal surface areas with low diffusional resistance in the gels. The surfaces of the gels were also presented (Fig. 1). The 'skin' of the gels can be seen, and the collapse of the surface pores may be due to the freeze-drying process.

INVESTIGATION OF FREE RADICAL CHEMISTRY OF THE IN VITRO MODEL SYSTEM

It is well established that HO. can be generated from a reaction known as the biologic Fenton reaction, and this reaction requires the presence of H_2O_2 [22-25]. The generation of HO. from the biologic Fenton reaction has been shown to be a critical factor in various ROSinduced oxidative stresses [24-26]. H_2O_2 and HO. might be related to apoptosis in atherosclerosis [30]. Godley et al. also reported that blue light induces mitochondrial DNA damage and cellular aging [31].

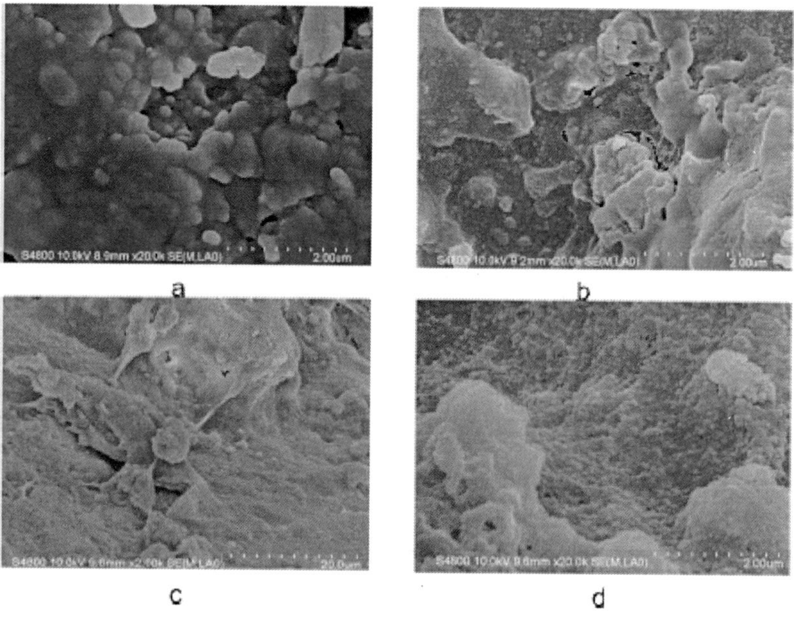

Figure 1. SEM images of the reactive surface of the composite under experimental conditions: a. Gel-2 and phenylalanine, b. Gel-3 and 2- deoxyguanosine, c. Gel-3 and phenylalanine, d. Gel-4 and 2′-deoxythymine.

The reactive nature of the surface has been investigated using the SEM and comparison confirms the reactive nature of the transformation. (Fig. 2).

CHEMISTRY OF REACTIVE OXYGEN SPECIES (ROS) UNDER "DENTAL CONDITIONS" IN VITRO

Enhanced oxidative damage after diverse stimuli has been confirmed to be an initial event in the development of several important diseases and adverse reactions [32-34]. The increased and uncontrollable production of oxygen centred free species in the specific and highly vascular region will increase detrimental effects including peroxidation of membranes and eventual cell death. In contrast, the oxidative insults can be partially prevented by some free radical quenchers such as Vitamin E, flavonoids and superoxide dismutase [35, 36].

Thus, antioxidants that react preferentially with reactive oxygen centred radicals to inactivate them might have therapeutic applications in treating free radical-induced cell damage and unnecessary cell death [35, 36]. Chitosan, the linear polymer of "-glucosamine in β (1,4) linkage, is a cellulose-like biopolymer present in the exoskeleton of crustaceans and in cell walls of fungi and insects [37]. Chitosan oligosaccharides (low-molecular weight chitosan, COS) are depolymerized products of chitosan and can be obtained by chemical and enzymatic hydrolysis of chitosan [38]. During the past decades, COS were known to have various biological activities [39-43] including its antioxidant capacity [44]. By using electron spin resonance (ESR) spintrapping technique, COS and their derivatives can show high scavenging activities on 1,1-diphenyl-2-picrylhydrazyl (DPPH), hydroxyl, superoxide and carbon-centered radicals [44-46]. Studies also showed that COS with different molecular weight (1-10 kDa) could inhibit free radical damages to murine melanoma cell line (B16F1) in a dose-dependent manner [47]. The link between oral diseases and free radical damage requires further investigation and is one of the main objectives of this investigation.

Control experiments established the integrity of the in vitro procedures to assess intracellular ROS levels. Under unphotolized conditions or exposure of reaction mixture and peroxide and absence of CuSO4, there was virtually no fluorescence. Light exposure generated levels of ROS that were 4-5 times higher than those induced by the diamide controls, this baseline data indicates the importance of finding molecular origin of these transformations and finding alternative ways of preventing this potentially detrimental reactions from occurring (Scheme 1).

It is well established that phenolic antioxidants are useful quenchers for excessive free radical generation with a potential application in disease chemoprevention. These are subsequently of particular interest in the development of the prototype model in alternative antioxidant delivery systems

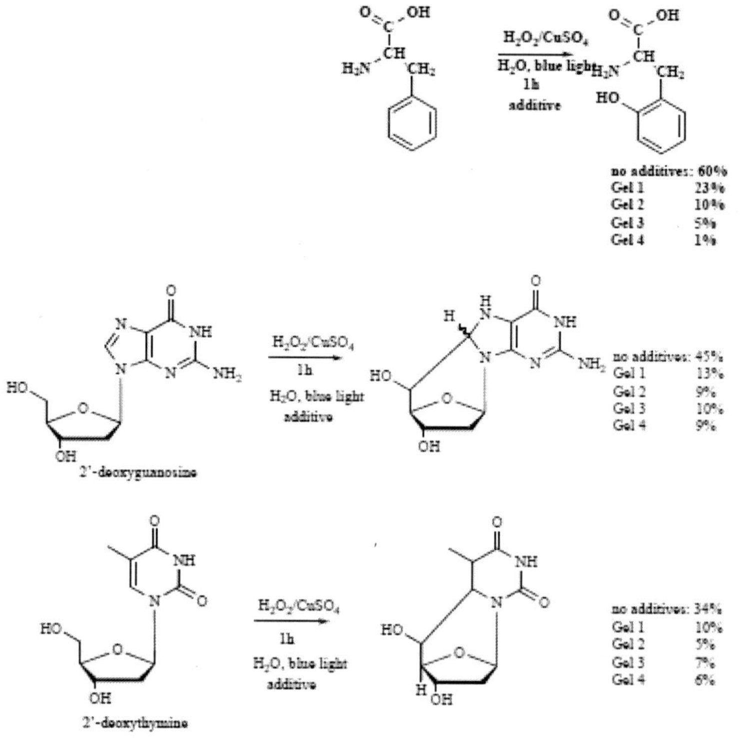

	no additives	Gel 1	Gel 2	Gel 3	Gel 4
Phenylalanine ($H_2O_2/CuSO_4$, H_2O, blue light, 1h, additive)	60%	23%	10%	5%	1%
2'-deoxyguanosine ($H_2O_2/CuSO_4$, 1h, H_2O, blue light, additive)	45%	13%	9%	10%	9%
2'-deoxythymine ($H_2O_2/CuSO_4$, 1h, H_2O, blue light, additive)	34%	10%	5%	7%	6%

Scheme 1. Reaction outcomes under Fenton conditions and potential biomarkers.

[48]. Extensive kinetic and mechanistic data exists on the formation and properties of the antioxidant, chitosan. Thus, the system is ideal for the assessment of performance of the complex in fundamental free radical transformations such as Fenton reaction and radical cascade reactions, which represent some of the key free

radical transformations in vivo responsible for DNA damage, disease state and eventual cell death [49, 50].

Several classical free radical reaction types such as Hydrogen Atom Transfer reaction, Carbon-Carbon bond formation via free radical mediated cascade reaction and free radical hydroxylation were investigated in artificial saliva using H_2O_2/Cu^{2+} and hypophosphorous acid/surfactant (H_3PO_2/Bu_4NBr) in water and as trial reactions probing the extent of oxidative stress and free radical damage (Scheme 2). Typical free radical precursors such as tertiary bromides of protected commonly used NSADs such as naproxen and ibuprofen were used in hydrogen atom transfer reactions. Racemic phenylalanine, 2-deoxythymine and 2- deoxyguanosine were used as typical precursors for the Fenton reactions. All reaction mixtures were analyzed using HPLC/MS and mixtures were further purified using conventional chromatographic techniques. For comparison purposes, reactions were conducted in the aqueous media with no inhibitor, chitosan Gel 1 and corresponding antioxidant under previously determined conditions for optimal generation of free radical transformation. Reactions in the presence of chitosan, antioxidant or chitosan: antioxidant gel demonstrated significant inhibition of the undesirable reaction outcome and effectiveness of the proposed methodology.

CONCLUSION

We have demonstrated the application of two effective molecular prototype delivery systems able to harness free radical reactivity within the laboratory where biological processes can be studied and controlled, in an attempt to translate the in vitro research for the development of new preventative measures for disease states mediated by free radicals.

Further studies are needed to elucidate more precisely the mechanisms of biological damage from oxidative stress and the mechanisms for defense of the dental pulp when using a wide variety of free radical generation procedures such as bleaching, blue light irradiation and LED light source commonly employed in dentistry. The positive influence of gels, containing chitosan/ antioxidant in the hydrogen transfer transformations broadens the potential application of the methodology for controlling free radical damage in the oral environment.

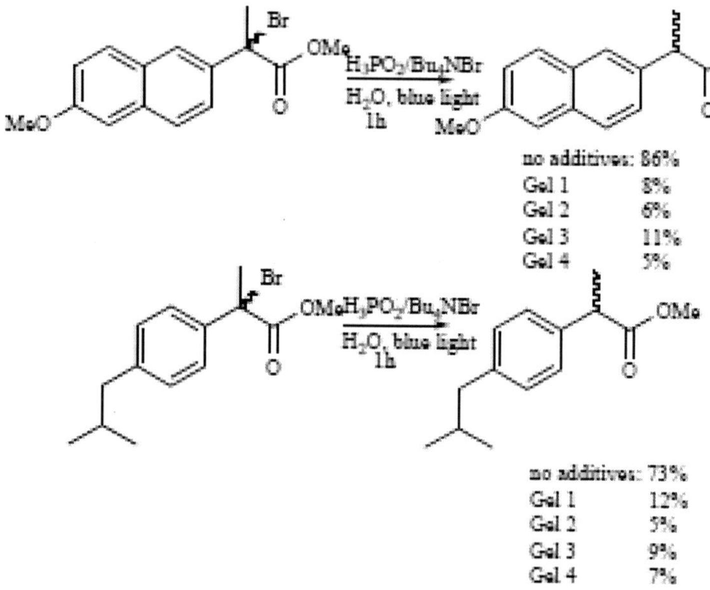
Br
OMe
H3PO2/Bu4NBr
H2O, blue light
1h
MeO
OMe
O
no additives: 86%
Gel 1 8%
Gel 2 6%
Gel 3 11%
Gel 4 5%
Br
OMe H3PO2/Bu4NBr
H2O, blue light
1h
OMe
no additives: 73%
Gel 1 12%
Gel 2 5%
Gel 3 9%
Gel 4 7%

Scheme 2. Reaction outcomes under Hydrogen Atom transfer conditions and potential pro-drug prototypes.

CONFLICT OF INTEREST

The authors confirm that this article content has no conflicts of interest.

ACKNOWLEDGEMENT

Declared none.

REFERENCES

1. Sandler, J.S.; Colin, P.L.; Kelly, M.; Fenical, W. Sterol mutants of Saccharomyces cerevisiae: chromatographic analyses. J. Org. Chem., 2006, 71, 8684.
2. Schmalz, G. Concepts in biocompatibility testing of dental restorative materials. Clin. Oral Invest., 1997, 1, 154.

3. Peumans, M.; Kanumilli, P.; De-Munck, J.; Van Landuyt, K.; Lambrechts, P.; Van Meerbeek, B. Clinical effectiveness of contemporary adhesives: a systematic review of current clinical trials. Dent. Mater., 2005, 21, 864.
4. Murray, P.E.; Windsor, L.J.; Hafez, A.A.; Stevenson, R.G.; Cox, C.F.; Comparison of pulp responses to resin composites. Oper. Dent., 2003, 28, 242.
5. Olivicr, A.; Grobler, S.R.; Osman, Y. Cytotoxicity of seven recent dentine bonding agents on mouse 3T3 fibroblast cells. Open J. Stomatol., 2012, 2, 244.
6. Chen, R.S.; Liu, C.C.; Tseng, W.Y.; Jeng, J.H.; Lin, C.P.; Cytotoxicity of three dentin bonding agents on human dental pulp cells. J. Dent., 2003, 31, 223.
7. Chen, R.S.; Liuiw, C.C.; Tseng, W.Y.; Hong, C.Y.; Hsieh, C.C.; Jeng, J.H. The effect of curing light intensity on the cytotoxicity of a dentin-bonding agent. Oper. Dent., 2001, 26, 505.
8. de Souza Costa, C.A.; Vaerten. M.A.; Edwards, C.A.; Hanks. C.T. Cytotoxic effects of current dental adhesive systems on immortalized odontoblast cell line MDPC-23. Dent. Mater., 1999, 15, 434.
9. Vajrabhaya, L.O.; Pasasuk, A., Harnirattisai, C., Cytotoxicity evaluation of single component dentin bonding agents. Oper. Dent., 2003, 28, 440.
10. Szep, S.; Kunke, l.A.; Ronge, K.; Heidemann, D. Cytotoxicity of modern dentin adhesives—in vitro testing on gingival fibroblasts. J. Biomed. Mater. Res., 2002, 63, 53.
11. Geurtsen, W.; Spahl, W.; Muller, K.; Leyhausen, G. Aqueous extracts from dentin adhesives contain cytotoxic chemicals. J. Biomed. Mater. Res., 1999, 48, 772.
12. Koliniotou-Koubia, E.; Dionysopoulos. P.; Koulaouzidou, E.A.; Kortsaris. A.H.; Papadogiannis, Y. In vitro cytotoxicity of six dentin bonding agents. J. Oral. Rehabil., 2001, 28, 971.
13. Kaga, M.; Noda, M.; Ferracane, J.L.; Nakamura, W.; Oguchi, H.; Sano, H.; The in vitro cytotoxicity of eluates from dentin bonding resins and their effect on tyrosine phosphorylation of L929 cells. Dent. Mater., 2001, 17, 333.
14. Mantellini, M.G.; Botero, T.M.; Yaman, P.; Dennison, J.B.; Hanks, C.T. Adhesive resin induces apoptosis and cell-cycle arrest of pulp cells. J. Dent. Res., 2003, 82, 592.
15. About, I.; Camps, J.; Burger, A.S.; Mitsiadis, T.A.; Butler, W.T.; Franquin, J.C. Polymerized bonding agents and differentiation in vitro of human pulp cells into odontoblast-like cells. Dent. Mater,. 2005, 21, 156.
16. Thonemann, B.; Schmalz, G. Immortalization of bovine dental papilla cells with simian virus 40 large t-antigen. Arch. Oral. Biol., 2000, 45, 857.
17. Thonemann, B.; Schmalz, G. Bovine dental papilla-derived cells immortalized with HPV 18 E6/E7. Eur. J. Oral. Sci., 2000, 108, 432.
18. Galler, K.M.; Schweikl, H.; Thonemann, B.; D'Souza, R.N.; Schmalz, G. Human pulp-derived cells immortalized with Simian Virus 40 T-antigen. Eur. J. Oral Sci., 2006,114, 138.

19. Schmalz, G.; Schuster, U.; Nuetzel, K.; Schweikl, H. An in vitro pulp chamber with three-dimensional cell cultures. J. Endodont., 1999, 25, 24-9.
20. Schuster, U.; Schmalz, G.; Thonemann, B.; Mendel, N.; Metzl, C. Cytotoxicity testing with three-dimensional cultures of transfected pulp-derived cells. J. Endodont., 2001; 27: 259-65.
21. Spagnuolo, G.; Annunziata, M.; Rengo, S. Cytotoxicity and oxidative stress caused by dental adhesive systems cured with halogen and LED lights. Clin. Oral Invest., 2004, 8, 81.
22. Atsumi, T.; Iwakura, I.; Fujisawa, S.; Ueha, T. The production of reactive oxygen species by irradiated camphorquinone-related photosensitizers and their effect on cytotoxicity. Arch. Oral Biol., 2001, 46, 391.
23. Spagnuolo, G.; D'Anto, V.; Cosentino, C.; Schmalz, G.; Schweikl, H.; Rengo, S. Effect of N-acetyl-l-cysteine on ROS production and cell death caused by HEMA in human primary gingival fibroblasts. Biomaterials, 2006, 27, 1803.
24. Janke, V.; von-Neuhoff, N.; Schlegelberger, B.; Leyhausen, G.;Geurtsen, W. TEGDMA causes apoptosis in primary human gingival fibroblasts. J. Dent. Res., 2003, 82, 814.
25. Schweikl, H.; Hartmann, A.; Hiller, K.A.; Spagnuolo, G.; Bolay, C.; Brockhoff, G. Inhibition of TEGDMA and HEMA-induced genotoxicity and cell cycle arrest by N-acetylcysteine. Dent. Mater., 2007, 23(6), 688.
26. Godley, B.F.; Shamsi, F.A.; Liang, F.Q.; Jarrett, S.G.; Davies, S.; Boulton, M. Blue light induces mitochondrial DNA damage and free radical production in epithelial cells. J. Biol. Chem., 2005, 280, 21066.
27. Jones, C.A.; Huberman, E.; Cunningham, M.L.; Peak, M.J.; Mutagenesis and cytotoxicity in human epithelial cells by far- and near-ultraviolet radiations: action spectra. Radiat. Res., 1987, 110, 244.
28. Peak, J.G.; Peak, M.J. Comparison of initial yields of DNA-toprotein crosslinks and single-strand breaks induced in cultured human cells by far- and nearultraviolet light, blue light and X-rays. Mutat. Res., 1991, 246,187.
29. Peak, J.G.; Peak, M.J. Induction of slowly developing alkali-labile sites in human P3 cell DNA by UVA and blue- and green-

light photons: action spectrum. Photochem. Photobiol., 1995, 61, 484.

30. Peak, J.G.; Peak, M.J.; Sikorski, R.S.; Jones, C. A. Induction of DNA-protein crosslinks in human cells by ultraviolet and visible radiations: action spectrum. Photochem. Photobiol., 1985, 41, 295.
31. Noell, W. K.; Walker, V. S.; Kang, B. S.; Berman, S. Retinal damage by light in rats. Invest. Ophthalmol., 1996, 5, 450.
32. Setlow, R.B.; Grist, E.; Thompson, K.; Woodhead, A.D. Wavelengths effective in induction of malignant melanoma. Proc. Natl. Acad. Sci. USA., 1993, 90, 6666.
33. Setlow, R.B.; Woodhead, A.D. Temporal changes in the incidence of malignant melanoma: explanation from action spectra. Mutat. Res., 1994, 307, 365-74.
34. Tyrrell, R.M.; P. Werfelli, P.; Moraes, E.C. Lethal action of ultraviolet and visible (blue-violet) radiations at defined wavelengths on human lymphoblastoid cells: action spectra and interaction sites. Photochem. Photobiol., 1984, 39, 183
35. Yoshida, A.; Yoshino, F.; Tsubata, M.; Ikeguchi, M.; Nakamura, T.; Lee, M. C. Direct assessment by electron spin resonance spectroscopy of the antioxidant effects of French maritime pine bark extract in the maxillofacial region of hairless mice. J. Clin. Biochem. Nutr., 2011, 49, 79.
36. Winkler, B.S.; Boulton, M.E.; Gottsch, J.D.; Sternberg, P. Oxidative damage and age-related macular degeneration. Mol. Vis., 1999, 5, 32.
37. Caughman, W.F.; Rueggeberg, F.A.; Curtis Jr. J.W.; Clinical guidelines for photocuring restorative resins. J. Am. Dent. Assoc., 1995, 126, 1280.
38. Marson, F.; Sensi, L.; Vieira, L.; Araújo, E. Clinical evaluation of in-office dental bleaching treatments with and without the use of light-activation sources, Oper. Dent., 2008, 33, 15.
39. Luk, K.; Tam, K.; Hubert, M. Effect of light energy on peroxide tooth bleaching. J. Am. Dent. Assoc., 2004, 135, 194.
40. Nomoto, R.; McCabe, J.F.; Hirano, S. Comparison of halogen, plasma and LED curing units. Oper. Dent., 2004, 29, 28.
41. Ohira, A.; Ueda, T.; Ohishi, K.; Hiramitsu, T.; Akeo, K.; Obara, Y.

Oxidative stress in ocular disease. Nihon Ganka Gakkai Zasshi., 2008,112, 22.

42. Byers, M.R. Dental sensory receptors. Int. Rev. Neurobiol., 1984, 25, 39.
43. Walker, J.W.; Martin, H.; Schmitt, F.R.; Barsotti, R.J. Rapid release of an alpha-adrenergic receptor ligand from photolabile analogues. Biochemistry., 1993, 32, 1338.
44. Hockberger, P.E.; Skimina, T.A.; Centonze, V.E.; Lavin, C.; Chu, S.; Dadras, S.; Reddy, J.K.; White, J.G. Activation of flavincontaining oxidases underlies lightinduced production of H2O2 in mammalian cells. Proc. Natl. Acad. Sci. USA., 1996, 96, 6255.
45. Peak, M.J.; Peak, J.G. Solar-ultraviolet-induced damage to DNA. Photodermatol., 1989, 6, 1.
46. Specht, S.; Leffak, M.; Darrow, R.M.; Organisciak, D.T. Damage to rat retinal DNA induced in vivo by visible light. Photochem.\ Photobiol., 1999, 69, 91.
47. Cunningham, M.L.; Krinsky, N.I.; Giovanazzi, S.M., Peak, M.J. Superoxide anion is generated from cellular metabolites by solar radiation and its components. J. Free Radic. Biol. Med., 1985, 1, 381.
48. Peak, M.J.; Peak, J.G. Hydroxyl radical quenching agents protect against DNA breakage caused by both 365-nm UVA and by gamma radiation. Photochem. Photobiol., 1990, 51, 649.
49. Bartold, P.M.; Wiebkin, O.W.; Thonard, J.C.; The effect of oxygen- derived free radicals on gingival proteoglycans and hyaluronic acid. J. Periodontal. Res., 1984, 19, 390.
50. Chapple, I.L. Role of free radicals and antioxidants in the pathogenesis of the inflammatory periodontal diseases. Clin. Mol. Pathol., 1996, 49, M247-55.

Chapter 6

RESPONSE SURFACE MODELING AND OPTIMIZATION OF ELECTROSPUN NANOFIBER MEMBRANES

M. Essalhi, M. Khayet*, C. Cojocaru, M.C. García-Payo, and P. Arribas

ABSTRACT

The experimental design and response surface methodology (RSM) have been used to develop predictive models for simulation and optimization of electrospun polyvinylidene fluoride non-woven membranes. The objective is to prepare electrospun fibers with small diameters and narrow diameter distribution. The factors considered for experimental design were the polymer dope solution flow rate, the applied electric voltage and the distance between the needle tip and the collector. A full factorial design was considered. The obtained electrospun fibers were characterized by scanning electron microscopy. The response for the model was the quality loss function that takes into account the quadratic effects ofboth the weighted arithmetic mean of the fibers diameter and the standard deviation. Minimal output response has been predicted and confirmed experimentally. The optimum operating conditions guarantying a small polyvinylidene fluoride nanofiber diameter with a narrow distribution were a voltage of 24.1 kV, an air gap of 27.7 cm and a polymer flow rate of1.23 mL/h. The fabricated optimum membrane was characterized by different techniques and applied for desalination by membrane distillation. The obtained permeate fluxes in this study are higher than those reported so far for electrospun nanofibrous membranes.

INTRODUCTION

Electrospinning has been recognized as an efficient technique for the fabrication of polymer nanofibers. These haveattracted increasing attentions in the last ten years because oftheir very large surface area to volume ratio, flexibility in surface functionalities and superior mechanical performance compared with any other known form of materials. These outstanding properties make the polymer nanofibers optimal candidates for many advanced applications in fields such as biomedical engineering and biotechnology, environmental engineering, energy storage, tissue engineering, drug delivery, affinity membranes, enzyme immobilization, etc [1-6].

Electrospinning can also organize nanofibers of various types such as porous, hollow and core/sheath into well defined arrays or hierarchical architectures in three dimensional networks. Numerous studies have been carried out to gain deep understanding of the process for a better control of fiber formation [7-14].Nowadays, systematic investigations of the effects of electrospinning variables on diameter and morphology of the electrospun fibers are of great interest. Obviously, there is an important need to produce fibers with small and uniform size so that the electrospinning process can be reproduced in large industrial applications [2, 5, 15]. Many parameters can affect the morphological structure and dimensions of electrospun fibers. These are system parameters such as polymer type and its molecular weight, polymer concentration, solvent type and polymer solution properties (viscosity, conductivity and surface tension); process parameters such as electric potential or voltage, flow rate of polymer solution, distance between the capillary and collector and ambient parameters (temperature, humidity and air velocity) [7-10, 16- 21]. Moreover, for preparation of nanofibrous membranes, the conventional or classical method of experimentation, which involves changing one of the independent parameters while maintaining the others fixed at given values, has been considered [16-21]. As it is well known, this conventional method of experimentation involves many tests, which are time-consuming, ignores interaction effects between the operating parameters and induces a low efficiency in optimization. These limitations can be avoided by applying the Response Surface Methodology (RSM) that involves statistical design of experiments (DoE) in which all factors are varied together over a set of experimental runs [22, 23]. In fact, the statistical method of experimental design offers several advantages over the frequently used conventional method being rapid and reliable, helps in understanding the interaction effects between factors and reduces the total number of experiments tremendously resulting in saving time and costs of experimentation. Moreover, RSM can be used to evaluate the relative significance of several affecting factors even in the presence of complex interactions [22-29].

In recent years various statistical experimental designs and RSM have been applied progressively to different processes [22-29]. However, among them few reports were dedicated to electrospinning [20, 30]. Yördem et al. [20] studied the effects of electrospinning parameters on polyacrylonitrile (PAN) nanofiber diameter using RSM. Their investigations were carried out using only two variables (applied voltage, solution concentration) but several collector distances. The effect of the applied voltage on fiber diameter was insignificant when the solution concentration and collector distance were high. Similarly, Gu et al. [30] applied RSM for PAN nanofibers and also reported no significant effect of the voltage on the PAN nanofibers. Both studies have been conducted considering two variables while the third parameter was maintained fixed, and therefore possible interactions between the three parameters were not studied [20, 30].

Polyvinylidene fluoride (PVDF) is an attractive material used in many applications due to its outstanding properties such as high mechanical strength, thermal stability, chemical resistance and good electrochemical stability compared to other commercialized polymeric materials. Electrospinning technique has also been applied to the fabrication of PVDF nanofibers and fibrous thin films for various applications [11-14].

In the present study a full factorial experimental design for fabrication of electrospun PVDF fibers has been considered. The polymer solution parameters (polymer type, molecular weight, solvents) and the environmental conditions (temperature and humidity) are maintained the same to prepare all ENMs. The main objective of this paper is to investigate the individual and mutual effects of the electrospinning variables (applied voltage, polymer solution flow rate and distance between the needle tip and the collector) on the diameter of the electrospun PVDF fibers as well as on fiber distribution. Furthermore, the optimum electrospinning conditions to ensure minimum fiber diameter with a narrow size distribution has been determined.

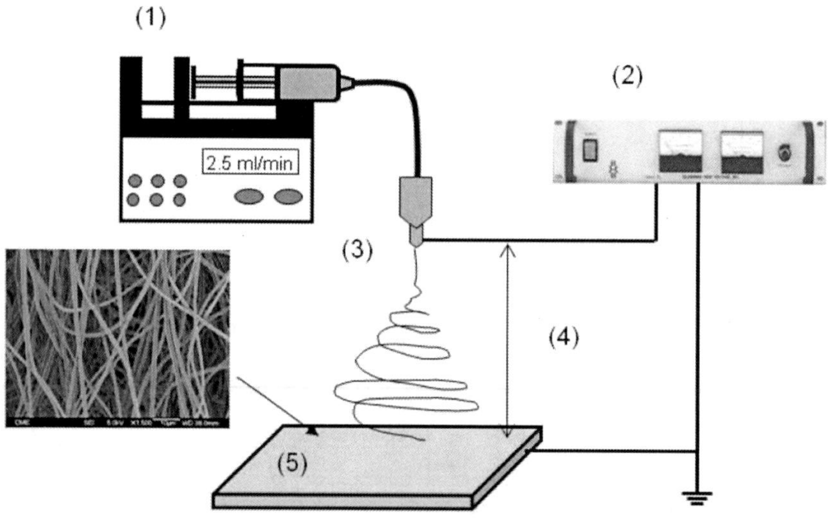

Figure 1. (1). Schematic diagram of electrospinning set-up. (1) Syringe with polymer solution, (2) high voltage supply, (3) spinneret, (4) distance between the needle tip and the collector, (5) collector.

An interesting application for ENMs is the non-isothermal distillation, which can be carried out for advanced water treatments without applying any transmembrane hydrostatic pressure and therefore self sustained webs can be used [31-35]. Therefore, the fabricated optimum membrane was characterized by different techniques and applied for desalination by direct contact membrane distillation (DCMD) using different salt (NaCl) aqueous concentrations. The DCMD performance is compared to other electrospun nanofibrous membranes [34, 35].

MATERIALS AND METHODOLOGY

Materials

The spinning solutions were prepared from the polymer PVDF (Mw = 275 kg/mol and Mn = 107 kg/mol) and the mixed solvents N,N-dimethyl acetamide (DMAC) and acetone purchased from Sigma-Aldrich Chemical Co. Isopropyl alcohol (IPA) was used to determine the void volume fraction and the size of the inter-fiber space, and the sodium chloride (NaCl) used in DCMD experiments was also purchased from Sigma Aldrich Chemical Co.

PREPARATION OF ELECTRO-SPUN PVDF FIBERS

The polymer solution was prepared using 25 wt% PVDF in the mixture 20 wt% acetone in DMAC. The electrospinning set-up is shown in Fig. (1) and consists of a syringe (50 ml, Nikepal) to hold the polymer solution, a pump (KDS Scientific, model 200), two electrodes (a metallic needle of 0.60 mm internal diameter and a grounded copper collector covered with aluminum foil) and a DC voltage supply in the kV range (Iseg, TCIP300 304p). The formed fibers were then dried in an oven at 80°C for 5 min (i.e. post-treatment).m PVDF electrospun fibers have been prepared following the experimental design conditions summarized in Table 1. The electrospinning parameters are the dope solution flow rate F (mL/h), the voltage U (kV) and the distance A (cm) between the needle tip and the collector, named hereafter as collector distance.

Table 1. Full Factorial Design for Electrospinning Experiments

Run	Voltage		Collector Distance		Polymer Flow Rate	
	x_1	U (kV)	x_2	A (cm)	x_3	F (mL/h)
B1	+1	24.5	+1	28.23	+1	3.28
B2	-1	10.5	+1	28.23	+1	3.28
B3	+1	24.5	-1	11.77	+1	3.28
B4	-1	10.5	-1	11.77	+1	3.28
B5	+1	24.5	+1	28.23	-1	1.22
B6	-1	10.5	+1	28.23	-1	1.22
B7	+1	24.5	-1	11.77	-1	1.22
B8	-1	10.5	-1	11.77	-1	1.22
B9	0	17.5	0	20	0	2.25
B10	0	17.5	0	20	0	2.25

CHARACTERIZATION

The surface of the non-woven electrospun PVDF membranes was examined by a field emission scanning electron microscope (FESEM, JEOL Model JSM-6330F). Micrographs from the SEM analysis were analyzed by UTHSCSA Image Tool 3.0 to determine the fiber diameter. For each sample more than 5 SEM images have been considered and the diameters of a total number of 100 fibers have been measured. Statistical analysis have been applied in order to determine the fiber size distribution and to estimate the arithmetic weighted mean of the fiber diameters and their dispersion (i.e. weighted standard deviation).

The electrospun nanofibrous PVDF membrane, prepared using the obtained optimum electrospinning conditions over a period of 3h30min, was characterized by different techniques to determine the liquid entry pressure (LEP) of distilled water and saline aqueous solutions of different concentrations (12 g/L, 30 g/L and 60 g/L), the mean size of the inter-fiber space (di) by the wet/dry flow method, the advancing water contact angle (!a) by a computerized optical system CAM100 (7.1 μL water drop), the void volume fraction («) from density measurements and the thickness (δ) by the micrometer Millitron Phywe (Mahr Feinprüf, type TYP1202IC). Details of the characterization techniques used are explained elsewhere [32]. Direct contact membrane distillation (DCMD) was carried out using the fabricated optimum electrospun PVDF membrane under different salt (NaCl) concentrations (0 g/L, 12 g/L, 30 g/L and 60 g/L), a feed temperature of 80°C and permeate temperature of 20°C and a stirring rate of both the feed and permeate of 500 rpm. The experimental system used is detailed in [36].

RESULTS AND DISCUSSIONS

The SEM images, together with their corresponding histograms showing the sizes of the electrospun fibers, are presented in Fig. (2). Differences exist between the SEM images of the samples depending on the electrospinning conditions. The best electrospun fibers have been obtained for the experimental run 5 (B5 in Fig. 2). The corresponding electrospinning values facilitate stretching of polymer solution along the distance between the needle and the collector and enhance the solvent evaporation leading to the formation of electrospun fibers with small diameters. The worst spinning conditions correspond to the experimental run 4 (B4 in Fig. 2), which involves the setting of factors at the opposite levels to those in experimental run 5. Such conditions seem to hinder solvent evaporation leading to fusion of fibers. As a result, more fiber-to-fiber contacts are formed.

Based on the statistical analysis of the histograms shown in Fig. (2), the weighted arithmetic mean (λ_w) of the fiber diameters and the corresponding weighted standard deviation (s_w) have been determined as follows [37]:

$$\lambda_w = \lambda_0 + \frac{h}{N}\sum_{j=1}^{m} u_j \cdot FC_j \tag{1}$$

$$s_w = \sqrt{\left(\frac{1}{N}\cdot\sum_{j=1}^{m}\left(u_j^2 \cdot FC_j\right) - \left(\frac{1}{N}\cdot\sum_{j=1}^{m}\left(u_j \cdot FC_j\right)\right)^2\right)\cdot h^2} \tag{2}$$

where m denotes the number of bins (disjoint categories), h is the bin width $h = (\lambda_{max}-\lambda_{min})/m$. FC is the frequency count, N is the number of samples in the statistical set (in our case N=100), λ_o is the dominant characteristic of the statistical set that corresponds to the highest peak, u is a variable defined as $u = (\lambda_c - \lambda_0)/h$ and λ_c is the bin characteristic (or bin center). Finally, the quality loss function (Y) that summarizes the quadratic effect of both weighted arithmetic mean and standard deviation as response for factorial modeling and optimization has been considered. This response is defined as follows [37, 38]:

$$Y = \lambda_w^2 + s_w^2 \quad (3)$$

In this case low Y value means good electrospinning process performance (i.e. low values of λ_w and s_w). Table 2 summarizes the obtained values of λ_w, s_w and Y determined according to the experimental design. In general, it was found that an increase of s_w is associated to λ_w.

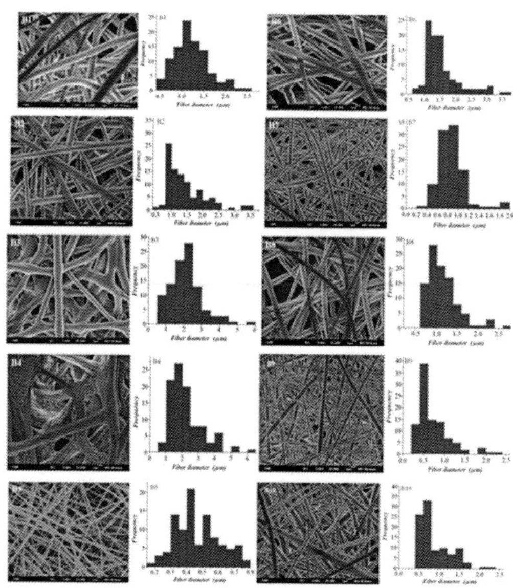

Figure. (2). SEM images of electrospun PVDF fibers (B1,B2,B3,...B10) prepared applying the electrospinning experimental runs summarized in Table 1.

Table 2. Responses Resulted from Statistical Analysis of Electrospun Fiber Diameter Distributions

Membrane	λ_w (µm)	s_w (µm)	$Y = \lambda_w^2 + s_w^2$ (µm²)
B1	1.218	0.410	1.652
B2	1.452	0.633	2.509
B3	2.130	0.928	5.398
B4	2.235	1.050	6.098
B5	0.470	0.138	0.240
B6	1.564	0.599	2.805
B7	0.874	0.285	0.845
B8	1.150	0.394	1.478
B9	0.738	0.413	0.715
B10	0.828	0.399	0.844

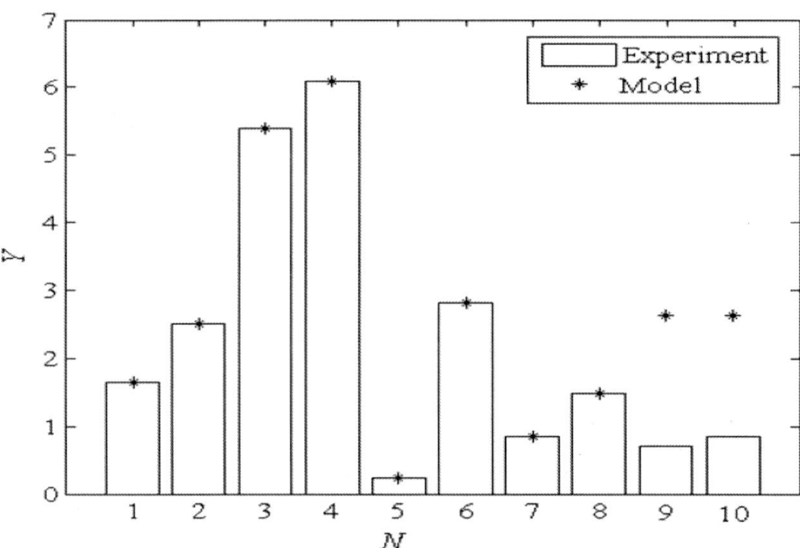

Figure. (3). Comparison between experimental and predicted data by the developed factorial model.

The electrospun fiber sample B5 exhibits the lowest values of λ_w, s_w and Y. In contrast, the electro-spun fiber sample B4 has the highest values of λ_w, s_w and Y.

Based on the regression techniques and the results presented in Tables 1 and 2, a factorial model with interactions has been developed. Eq. (4) shows the obtained factorial model in terms of coded variables:

$$\hat{Y} = 2.628 - 0.594x_1 - 0.827x_2 + 1.286x_3$$
$$-0.261x_1x_2 + 0.205x_1x_3 - 1.007x_2x_3 + 0.222x_1x_2x_3$$
$$\text{subjected to: } -1 \le x_i \le +1;\ \forall i = \overline{1,\ 3} \qquad (4)$$

where $\hat{Y}$ is the predictor of the response (quality loss function,Y). The significance of each individual regression coefficient has been tested by means of Student's t-test [39]. The analysis of variance (ANOVA) has been used to check the statistical significance of the factorial model. F-value has been determined based on the ratio of the mean square of group variance due to the error [40]. The larger is the difference of F-value from unity, the more certain it is that the designed variables (factors) adequately explain the variation in the mean of the data. In this case, the F-value is higher than 1 (2.159) and the coefficient of multiple determination R^2 indicated that 81.2% of the data variation can be explained by the factorial model. Therefore, the developed interaction factorial model can be accepted for the prediction of the response in the considered range of experimentation (valid region). It must be pointed out that the obtained regression coefficients in Eq. (4) can not be considered for electrospun modeling of other polymer solutions and other environmental conditions (i.e. temperature and humidity). The same DoE and RSM can be applied and other regression coefficients may be obtained.

Fig. (3) reports a comparison of the predicted and experimentally measured response. The predicted data are almost identical to the experimental ones for the orthogonal points (i.e runs 1-8). However, for the center point (i.e. runs 9 and 10) the discrepancy between the predicted and experimental data is higher compared to the other experimental runs. This means that the regression equation does not describe very accurately the response in the center point. This behavior can be attributed to the orthogonal property of the factorial design. However, based on the ANOVA statistical test the overall prediction may be considered satisfactory.

For graphical representation and analysis of response surface, the factorial model in terms of coded variables has been converted to an empirical model in terms of actual variables. The obtained factorial model in terms of actual variables is as follows:

$$\hat{Y} = -5.445 + 0.11U + 0.393A + 4.436F - 0.013U\,A$$
$$-0.046U\,F - 0.184A\,F + 3.741\times10^{-3}U\,A\,F \quad (5)$$

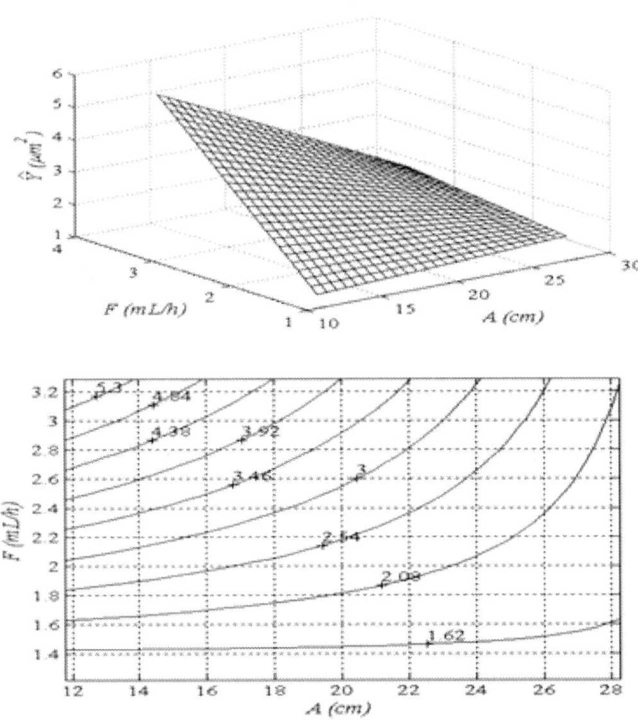

Figure. (4). Quality loss function (Y) versus the variables F (mL/h) and A (cm) maintaining U at 17.5 kV.

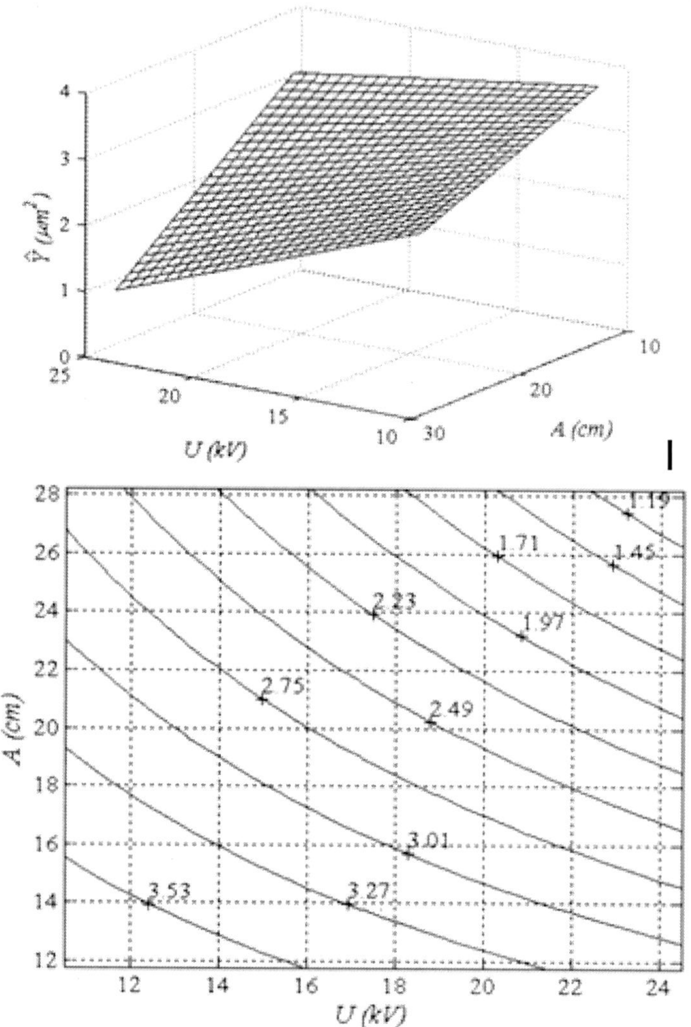

Figure. (5). Quality loss function (Y) versus the variables U (kV) and A (cm) maintaining F at 2.25 mL/h.

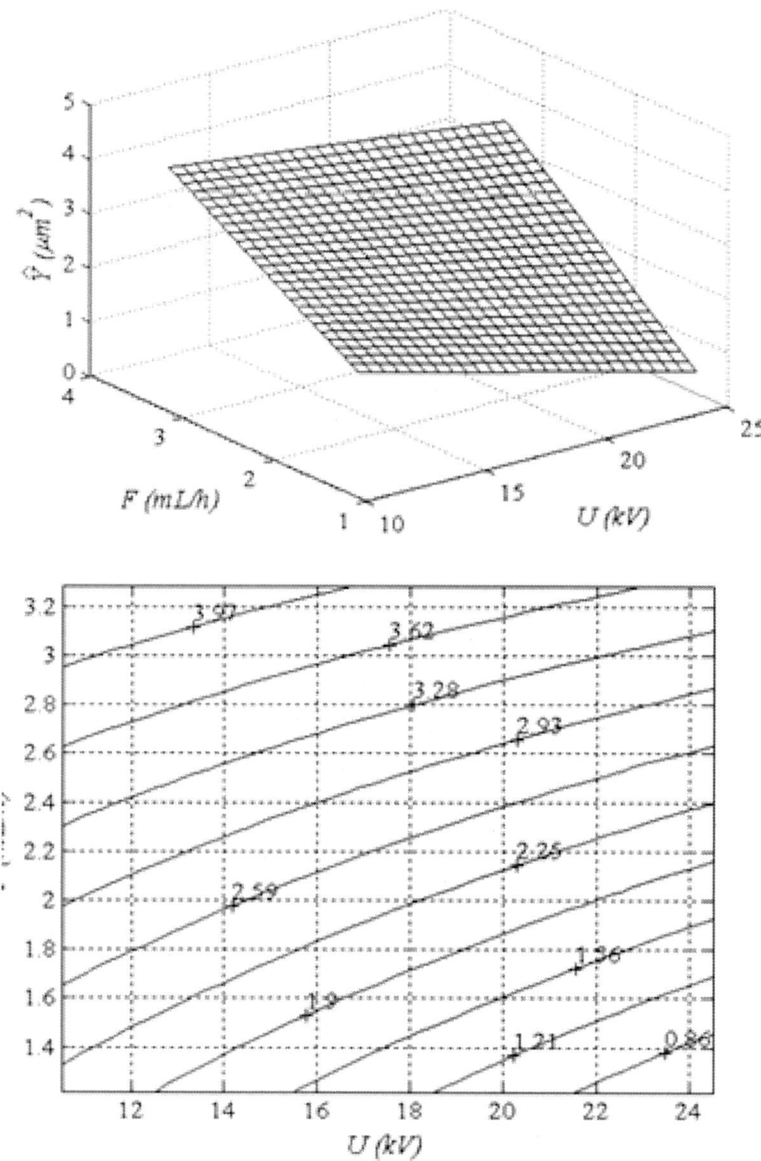

Figure. (6). Quality loss function (Y) versus the variables U (kV) and F (mL/h) maintaining A at 20 cm.

This model equation is valid for the following region of experimentation:

$$10.5 \leq U \leq 24.5 \text{ (kV)};\ 11.77 \leq A \leq 28.23 \text{ (cm)};$$

$$1.22 \leq F \leq 3.28 \text{ (mL/h)};$$

Figs. (4-6) present the response surface plots and contour lines maps of Y as a function of the design variables. It is known that polymer flow rate determines the quantity of the solution available for electrospinning.

Table 3. Electrospinning Optimal Point Determined by Monte Carlo Method

Voltage		Collector Distance		Polymer Flow Rate			
x_1	U (kV)	x_2	A (cm)	x_3	F (mL/h)	$Y_{predicted}$ (μm²)	$Y_{experimental}$ (μm²)
0.952	24.1	0.937	27.7	-0.989	1.23	0.328	0.314±0.099

When the polymer flow rate is increased, the diameter of the electrospun fiber is also increased [18, 21]. If the polymer flow rate is too high, greater volume of polymer solution will be drawn from the needle tip and the electrospinning jet will take more time to dry. As a result, the solvents in the deposited fibers over the collector may not have enough time to evaporate. Therefore, the residual solvent may cause the fibers to fuse together forming denser fibrous membrane. This will affect the volume charge density and the electrical current of the polymer solution, which increase or decrease depending on the polymer solution. Nasir *et al.* [18] observed that the PVDF fiber diameter decreased with increasing polymer flow rate up to 5 μL/min and then remained constant for higher flow rates.

The flight time of the electrospinning *jet along* the gap distance may affect considerably the fiber's characteristics. Decreasing the gas distance has the same effect as increasing the electrical voltage inducing higher electric field strength. When the gap distance is too short, the instability of the jet increases and the spinning solution cannot be fully stretched, resulting in greater fiber diameter. When the gap distance is too large, the strength of the electric field becomes weak resulting in an increase of fiber diameter and sometimes electrospinning is hard to accomplish. Depending on the polymer solution parameters, varying the distance may or may not have a significant effect on the fiber morphology. Nasir *et al.* [18] reported that the gap distance had no significant effect on the PVDF fiber diameter and explained that the increase of the gap distance induced a decrease of the electrical field strength when a constant electrical voltage was applied, whereas the solvent evaporation time of the polymer jet increased. Megelski *et al.* [21] also observed no significant change of the electrospun polystyrene fiber size with the change of the gap distance. However, inhomogeneous distribution of elongated beads took place when the gap distance was reduced. Park *et al.* [41] observed a decrease of the diameter of electrospun polyvinylacetate (PVAc) fiber with increasing the gap distance down to a minimum value followed by a gradual increase of the fiber diameter. This is due to the decrease in the electrostatic field strength resulting in less stretching of the fibers and indicates that there is an optimal electrostatic field strength below which the stretching of the solution will decrease resulting in increased fiber diameter. Therefore, the study of interaction effect in electrospinning is of great interest.

Fig. (4) shows the influence of the polymer flow rate F (mL/h) and the collector distance A (cm) on Y. As can be observed a strong interaction effect exists between these two parameters F and A. The decrease of the flow rate reduces Y, and due to the mutual interaction between F and A the overall effect of F is more apparent at lower level of A. On the contrary, the decrease of the collector distance leads to an enhancement of the quality loss function. Owing to the interaction effect, the influence of A is tiny at lower F and very strong at higher values of F. A high collector distance and low flow rate minimize the quality loss function and improve the performance of the electrospinning process. This can be attributed to the fact that such setting of factors ensures a sufficient time for solvent evaporation. It is worth quoting that the high voltage will produce the necessary charges on the polymer solution initiating electrospinning process when the electrostatic force in the solution overcomes the surface tension of the solution. When the applied voltage is higher, the greater amount of the induced charges will cause faster acceleration of the electrospinning jet and then a higher quantity of polymer solution will be drawn from the needle tip. These will result in a larger fiber diameter. Depending on the polymer flow rate of the dope and the polymer concentration, a high voltage may be required so that the Taylor cone is stable. The columbic repulsive force in the jet will then stretch the viscoelastic solution. In various cases, a higher electric voltage causes greater stretching of the polymer solution reducing in this way the diameter of electrospun fibers.

Fig. (5) illustrates the effects of U and the collector distance on Y. As can be seen, the increment of both variables diminishes the response and improves the electrospinning process performance. The effect of U is stronger at higher levels of A, and the effect of A is more evident at high values of U. Therefore, according to the predictions, the best result must be obtained for high values of both the applied voltage and the collector distance. The effects of U and F on Y are plotted in Fig. (6). The graphical analysis reveals that increasing U and decreasing F reduce Y. The interaction effects between the applied voltage and the polymer flow rate is minor compared to the previous ones (Figs. 4 and 5). However, the influence of U is stronger at lower levels of F. In contrast, the overall effect of F is stronger for higher levels of U. According to the response surface plot shown in Fig. (6) the smallest fiber diameters are obtained applying high values of the U and low values of F. To determine the optimum electrospinning conditions, the factorial model (Eq. 4) has been used. Monte Carlo method was employed for stochastic simulations and optimization in order to minimize the objective function. Table 3 reports the obtained optimal solution in terms of both coded and actual variables. Experimental confirmation run was performed using the optimum electrospinning conditions in order to confirm or disapprove the optimal point from experimental standpoint. SEM image and histogram of the electrospun PVDF fiber prepared applying the determined optimum experimental conditions are shown in Fig. (7). It was found 4.3 % deviation between the predicted quality loss function and the experimental one confirming the optimal point.

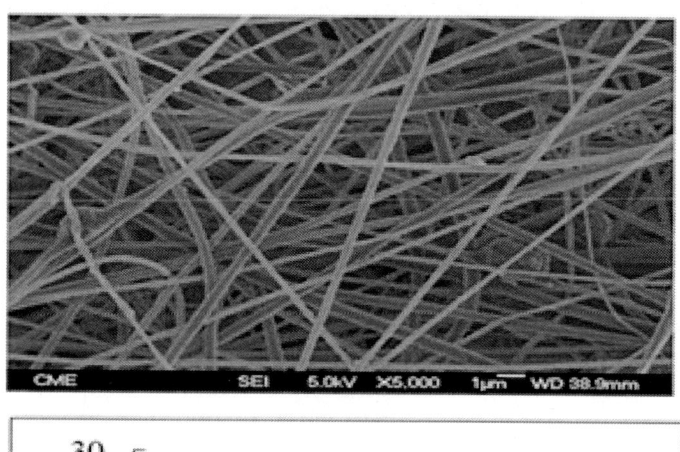

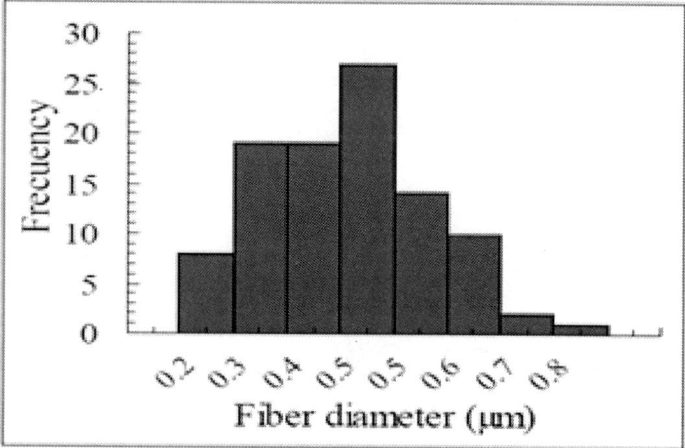

Figure. (7). SEM image and fiber diameter distribution of the electrospun

PVDF fiber prepared applying the optimum experimental conditions. By applying the obtained optimum electrospinning conditions, a PVDF electrospun membrane was prepared during 3h30min electrospinning time and characterized by different techniques as indicated previously. The results are summarized in Table 4. The water contact angle of the prepared optimum PVDF nanofibrous membrane is greater than that of the PVDF nanofibrous membranes reported by Feng *et al.* [34] and Prince et al. [35] (128°). The obtained high water contact angle in this study is attributed to the distinct PVDF polymer used and to the small fiber diameter achieved of the PVDF electrospun membrane fabricated by optimum electrospinning conditions. Feng et al. [34] and Prince *et al.* [35] reported a higher LEP values for PVDF electrospun membranes than the LEP values obtained in this study, 121.4 kPa and 90 kPa, respectively. These results are due to the distinct PVDF polymer solution used, to the different electrospinning parameters applied and to the different maximum inter-fiber space. For instance, the optimum PVDF electrospun membrane prepared in this study exhibits a higher inter-fiber space (d_i) than those of the PVDF nanofibrous membranes prepared by Feng et al. [34] (0.32 μm) and Prince *et al.* [35] (0.58-0.64 μm). Moreover, the void volume fraction (ε) of the optimum PVDF electrospun membrane is found to be slightly higher than that of the PVDF electrospun membranes prepared by Feng et al. [34] (76%) and Prince *et al.* [35] (81%). As stated previously, the prepared PVDF membrane was applied for desalination by direct contact membrane distillation (DCMD) using different feed salt (NaCl) concentrations (distilled water, 12 g/L, 30 g/L and 60 g/L), 80°C feed temperature and 20°C permeate temperature. The obtained permeate fluxes were 58.8±0.2 kg/m2.h, 57.3±0.4 kg/m2.h, 53.5±0.4 kg/m2.h and 51.3±0.3 kg/m2.h, for distilled water, 12 g/L, 30 g/L and 60 g/L, respectively. The salt rejection factors were greater than 99.94%. The observed decrease of the permeate flux with the increase of the feed salt concentration is due to the reduction of the water vapour pressure at the feed/membrane interface, which decrease the driving force (i.e. transmembrane water vapour pressure), and to the concentration polarization effect [32]. Although the thickness of the optimum electrospun membrane prepared in this study is higher than that of the PVDF nanofibrous membranes prepared by Feng *et al.* [34] and Prince *et al.* [35], the obtained permeate fluxes are more than 4.4 times greater.

The highest permeate flux obtained by Feng *et al.* [34] for a PVDF electrospun membrane was 11.5 kg/m^2.h whereas that reported by Prince *et al.* [35] was even lower 5.8 kg/m^2.h.m The obtained high permeation flux in the present study may be attributed to the higher size of the inter-fiber space and to the smaller fiber diameter affecting to some extent the mechanism of mass transport through the inter-fiber space of the electrospun nanofibrous membranes [32, 33].

CONCLUSIONS

In this work a full factorial design 23 was employed for fabrication of electro-spun PVDF fibers. Three independent variables were considered for the first time in experimental design related to electro-spinning. These variables are the applied voltage, the polymer solution flow rate and the distance between the needle tip and the collector. As response of interest the quality loss function was used. This takes into account both the weighted arithmetic mean of fibers diameter and its dispersion. The main and interaction effects of the electro-spinning variables on experimental response wererevealed.

Table 4. Characteristics of the PVDF Nanofibrous Membrane Prepared Applying the Optimum Electrospinning Conditions: Void Volume Fraction (ε), Advancing Water Contact Angle (θ_a) Mean Size of the Inter-Fiber Space (d*i*), Liquid Entry Pressure (LEP) of Distilled Water and NaCl Aqueous Solutions (12 g/L, 30 g/L and 60 g/L)

ε (%)	θ_a (°)	δ (μm)	d_i (μm)	*LEP* (kPa)			
				Distilled Water	12 g/L	30 g/L	60 g/L
81.6 ±4.2	150.1 ±1.1	567.2 ±25.4	0.82 ±0.09	33.5 ±0.7	37.0 ±1.4	40.5 ±3.5	42.5 ±4.9

Thus, a strong interaction effect was detected between polymer flow rate and the collector distance by means of three dimensional surface plot and contour-line map. A high collector distance and a low flow rate both minimize the quality loss function and improve the performance of the electro-spinning process because both ensure a sufficient time for solvent evaporation through the air gap of the polymer jet. Finally, the optimal point was determined using the factorial model and Monte Carlo optimization method. Under the obtained optimum operating conditions, 1.23 mL/h polymer flow rate, 24.1 kV voltage and 27.7 cm air gap a small PVDF nano-fiber diameter and narrow dispersion were obtained experimentally.

The fabricated membrane applying the determined optimum electrospinning parameters was characterized by different techniques and applied for desalination by direct contact membrane distillation (DCMD). The obtained permeate fluxes were 58.8±0.2 kg/m2.h, 57.3±0.4 kg/m2.h, 53.5±0.4 kg/m2.h and 51.3±0.3 kg/m2.h, for distilled water, 12 g/L, 30 g/L and 60 g/L salt (NaCl) aqueous solutions, respectively; with salt rejection factors greater than 99.94%. These permeate fluxes are more than 4.4 times greater than those reported so far for electrospun nanofibrous membranes used in membrane distillation (MD).

The statistical experimental design and response surface methodology can be applied for other polymer solutions and other electrospinning environmental conditions (Temperature and humidity). Different regression coefficients (Eq. 4) and different optimum electrospinning conditions may be obtained.

CONFLICT OF INTEREST

The authors confirm that this article content has no conflicts of interest.

ACKNOWLEDGEMENTS

The authors gratefully acknowledge the financial support of the I+D+I project MAT2010-19249 (Spanish Ministry of Science and Innovation). M. Essalhi is thankful to the Middle East Desalination Research Centre (MEDRC, Project 06- AS-02).

REFERENCES

1. Li, D.; Xia, Y. Electrospinning of nanofibers: reinventing the wheel. Adv. Mater., 2004, 16, 1151.
2. Ramakrishna, S.; Fujihara, K.; Teo, W.E.; Yong, T.; Ma, Z.; Ramaseshan, R. Electrospun nanofibers: solving global issues. Mater. Today, 2006, 9, 40.
3. Bhardwaj, N.; Kundu, S.C. Electrospinning: A fascinating fiber fabrication technique. Biotechnol. Adv., 2010, 28, 325.
4. Huang, Z.M.; Zhang, Y.Z.; Kotaki, M.; Ramakrishna, S. A review on polymer nanofibers by electrospinning and their applications in nanocomposites. Compos. Sci. Technol., 2003, 63, 2223.
5. Liang, D.; Hsiao, B.S.; Chu, B. Functional electrospun nanofibrous scaffolds for biomedical applications. Adv. Drug Deliv. Rev., 2007, 59, 1392.

6. Jang, J.H.; Castano, O.; Kim, H.W. Electrospun materials as potential platforms for bone tissue engineering. Adv. Drug Deliv. Rev., 2009, 61, 1065.
7. Baji, A.; Mai, Y.W.; Wong, S.C.; Abtahi, M.; Chen, P.Electrospinning of polymer nanofibers: Effects on oriented morphology, structures and tensile properties. Compos. Sci. Technol., 2010, 70, 703.
8. Casper, C.; Stephens, J.; Tassi, N.; Chase, D.; Rabolt, J. Controlling surface morphology of electrospun polystyrene fibers: effect of humidity and molecular weight in the electrospinning process. Macromolecules, 2004, 37, 573.
9. Zhang, D.; Chang, J. Electrospinning of three-dimensional nanofibrous tubes with controllable architectures. Nano Lett., 2008, 8, 3283.
10. Jiang, H.L.; Fang, D.F.; Hsiao, B.S.; Chu, B.; Chen, W.L. Optimization and characterization of dextran membranes prepared by electrospinning. Biomacromolecules, 2004, 5, 326.
11. Yee, W.A.; Kotaki, M.; Liu, Y.; Lu, X. Morphology, polymorphism behavior and molecular orientation of electrospun poly(vinylidene fluoride) fibers. Polymer, 2007, 48, 512.
12. Rietveld, I.B.; Kobayashi, K.; Yamada, H.; Matsushige, K. Morphology control of poly(vinylidene fluoride) thin film made with electrospray. J. Colloid Interface Sci., 2006, 298, 639.
13. Choi, S.W.; Jo, S.M.; Lee, W.S.; Kim, Y.R. An electrospun poly (vinylidene fluoride) nanofibrous membrane and its battery applications. Adv. Mater. 2003, 15, 2027.
14. Zhao, Z.Z.; Li, J.Q.; Yuan, X.Y.; Li, X.; Zhang, Y.Y.; Sheng, J. Preparation and properties of electrospun poly (vinylidene fluoride) membranes. J. Appl. Polym. Sci., 2005, 97, 466.
15. Haghi, A.K.; Akbari, M. Trends in electrospinning of natural nanofibers. Phys. Status Solidi 2007, 204, 1830.
16. Deitzel, J.M.; Kleinmeyer, J.; Harris, D.; Tan, N.C.B. The effect of processing variables on the morphology of electrospun nanofibers and textiles. Polymer, 2001, 42, 261.
17. Ki, C.S. ;Baek, D.H.; Gang, K.D.; Lee, K.H.; Um, I.C.; Park, Y.H. Characterization of gelatin nanofiber prepared from gelatin-formic acid solution. Polymer, 2005, 46, 5094.
18. Nasir, M.; Matsumoto, H.; Danno, T.; Minagawa, M.; Irisawa, T.; Shioya, M.; Tanioka, A. Control of diameter, morphology, and structure of PVDF nanofiber fabricated by electrospray deposition. J. Polym. Sci. Part B: Polym. Phys., 2006, 44, 779.
19. Reneker, D.H.; Yarin, A.L. Electrospinning jets and polymer nanofibers. Polymer 2008, 49, 2387.
20. Yördem, O.S.; Papila, M.; Menceloğlu, Y.Z. Effects of electrospinning parameters on polyacrylonitrile nanofiber diameter: an investigation by response surface methodology. Mater. Design, 2008, 29, 34.

21. Megelski, S.; Stephens, J.S.; Chase, D.B.; Rabolt, J.F. Micro-and nanostructured surface morphology on electrospun polymer fibers. Macromolecules, 2002, 35, 8456.

22. Donglai, W.; Zhenshan, C.; Jun, C. Optimization and tolerance prediction of sheet metal forming process using response surface model. Computational Mater. Sci., 2008, 42, 228.

23. Khayet, M. ;Cojocaru, C.; Zakrzewska-Trznadel, G. Response surface modelling and optimization in pervaporation. J. Membr. Sci., 2008, 321, 272.

24. Idris, A.; Kormin, F.; Noordin, M.Y. Application of response surface methodology in describing the performance of thin film composite membrane. Sep. Purif. Technol., 2006, 49, 271.

25. Khayet, M.; Cojocaru, C.; García-Payo, M.C. Application of response surface methodology and experimental design in direct contact membrane distillation. Ind. Eng. Chem. Res., 2007, 46, 5673.

26. Ismail, A.F.; Lai, P.Y. Development of defect-free asymmetric polysulfone membranes for gas separation using response surface methodology. Sep. Purif. Technol., 2004, 40, 191.

27. Khayet, M.; Cojocaru, C.; García-Payo, M.C. Experimental design and optimization of asymmetric flat-sheet membranes prepared for direct contact membrane distillation. J. Membr. Sci., 2010, 351, 234.

28. Khayet, M.; Abu Seman, M.N.; Hilal, N. Response surface modeling and optimization of composite nanofiltration modified membranes. J. Membr. Sci., 2010, 349, 113.

29. Cui, W.; Li, X.; Zhou, S.; Weng, J. Investigation on process parameters of electrospinning system through orthogonal experimental design. J. Appl. Polym. Sci., 2007, 103, 3105.

30. Gu, S.Y.; Ren, J.; Vansco, G.J. Process optimization and empirical modeling for electrospun polyecrylonitrile (PAN) nanofiber precursor of carbon nanofibers. Eur. Polym. J., 2005, 41, 2559.

31. Khayet, M.; García-Payo, M.C. Nanostructured flat membranes for direct contact membrane distillation, PCT/ES2011/000091, WO/2011/117443, 2011.

32. Khayet, M.; Matsuura T. Membrane distillation: Principles and applications, Elsevier: The Netherlands, 2011.

33. Khayet, M. Membranes and theoretical modeling of membrane distillation: a review. Adv. Colloid Interface Sci., 2011, 164(1-2), 56.

34. Feng, C.; Khulbe, K.C.; Matsuura, T.; Gopal, R.; Kaur, S.; Ramakrishna, S.; Khayet, M. Production of drinking water from saline water by air-gap membrane distillation using polyvinylidene fluoride nanofiber membrane. J. Membr. Sci., 2008, 311, 1.

35. Prince, J.A.; Singh, G.; Rana, D.; Matsuura, T.; Anbharasi, V.; Shanmugasundaram, T.S. Preparation and characterization of highly hydrophobic poly(vinylidene fluoride)-clay nanocomposite nanofiber

membranes (PVDF-clay NNMs) for desalination using direct contact membrane distillation. J. Membr. Sci., 2012, 397- 398, 80.

36. Essalhi, M.; Khayet, M. Surface segregation of fluorinated modifying macromolecule for hydrophobic/hydrophilic membrane preparation and application in air gap and direct contact membrane distillation. J. Membr. Sci., 2012, 417-418, 163.

37. Taloi, D.; Bratu, C.; Florian, E.; Berceanu, E. Optimization of the metallurgical processes, (in Romanian), Didactica & Pedagogica Publisher: Bucharest, 1983.

38. Alexis, J. Taguchi method applied for industrial practice: Experimental designs, (in Romanian), Tehnica Publisher: Bucharest, 1999.

39. Akhnazarova, S.; Kafarov, V. Experiment optimization in chemistry and chemical engineering, Mir Publisher: Moscow, 1982.

40. Montgomery, D.C. Design and Analysis of Experiments, 5th ed.; Wiley & Sons: New York, 2001.

41. Park, J.Y.; Lee, I.H.; Bea, G.N. Optimization of the electrospinning conditions for preparation of nanofibers from polyvinylacetate (PVAc) in ethanol solvent. J. Ind. Eng. Chem., 2008, 14, 707.

Chapter 7

POLYMER NANOTECHNOLOGY: NANOCOMPOSITES

D.R. Paul[a, 1], L.M. Robeson[b],

[a] Department of Chemical Engineering and Texas Materials Institute, University of Texas at Austin, Austin, TX 78712, United States

[b] Lehigh University, 1801 Mill Creek Road, Macungie, PA 18062, United States

ABSTRACT

In the large field of nanotechnology, polymer matrix based nanocomposites have become a prominent area of current research and development. Exfoliated clay-based nanocomposites have dominated the polymer literature but there are a large number of other significant areas of current and emerging interest. This review will detail the technology involved with exfoliated clay-based nanocomposites and also include other important areas including barrier properties, flammability resistance, biomedical applications, electrical/electronic/optoelectronic applications and fuel cell interests. The important question of the "nano-effect" of nanoparticle or fiber inclusion relative to their larger scale counterparts is addressed relative to crystallization and glass transition behavior. Of course, other polymer (and composite)-based properties derive benefits from nanoscale filler or fiber addition and these are addressed.

INTRODUCTION

The field of nanotechnology is one of the most popular areas for current research and development in basically all technical disciplines. This obviously includes polymer science and technology and even in this field the investigations cover a broad range of topics. This would include microelectronics (which could now be referred to as nanoelectronics) as the critical dimension scale for modern devices is now below 100 nm. Other areas include polymer-based biomaterials, nanoparticle drug delivery, miniemulsion particles, fuel cell electrode polymer bound catalysts, layer-by-layer self-assembled polymer films, electrospun nanofibers, imprint lithography, polymer blends and nanocomposites. Even in the field of nanocomposites, many diverse topics exist including composite reinforcement, barrier properties, flame resistance, electro-optical properties, cosmetic applications, bactericidal properties. Nanotechnology is not new to polymer science as prior studies before the age of nanotechnology involved nanoscale dimensions but were not specifically referred to as nanotechnology until recently. Phase separated polymer blends often achieve nanoscale phase dimensions; block copolymer domain morphology is usually at the nanoscale level; asymmetric membranes often have nanoscale void structure, miniemulsion particles are below 100 nm; and interfacial phenomena in blends and composites involve nanoscale dimensions. Even with nanocomposites, carbon black reinforcement of elastomers, colloidal silica modification and even naturally occurring fiber (e.g., asbestos-nanoscale fiber diameter) reinforcement are subjects that have been investigated for decades. Almost lost in the present nanocomposite discussions are the organic–inorganic nanocomposites based on sol–gel chemistry which have been investigated for several decades [1], [2] and [3]. In essence, the nanoscale of dimensions is the transition zone between the macro-level and the molecular level. Recent interest in polymer matrix based nanocomposites has emerged initially with interesting observations involving exfoliated clay and more recent studies with carbon nanotubes, carbon nanofibers, exfoliated graphite (graphene), nanocrystalline metals and a host of additional nanoscale inorganic filler or fiber modifications.

This review will discuss polymer matrix based nanocomposites with exfoliated clay being one of the key modifications. While the reinforcement aspects of nanocomposites are the primary area of interest, a number of other properties and potential applications are important including barrier properties, flammability resistance, electrical/electronic properties, membrane properties, polymer blend compatibilization. An important consideration in this review involves the comparison of properties of nanoscale dimensions relative to larger scale dimensions. The synergistic advantage of nanoscale dimensions ("nano-effect") relative to larger scale modification is an important consideration. Understanding the property changes as the particle (or fiber) dimensions decrease to the nanoscale level is important to optimize the resultant nanocomposite. As will be noted, many nanocomposite systems noted in the literature can still be modeled using continuum models where absolute size is not important since only shape and volume fraction loading are necessary to predict properties. Nanoscale is considered where the dimensions of the particle, platelet or fiber modification are in the range of 1–100 nm. With the platelet or fiber, the smallest dimension is considered for that range (platelet thickness or fiber diameter).

FUNDAMENTAL CONSIDERATIONS

In the area of nanotechnology, polymer matrix based nanocomposites have generated a significant amount of attention in the recent literature. This area emerged with the recognition that exfoliated clays could yield significant mechanical property advantages as a modification of polymeric systems [4], [5] and [6]. The achieved results were at least initially viewed as unexpected ("nano-effect") offering improved properties over that expected from continuum mechanics predictions. More recent results have, however, indicated that while the property profile is interesting, the clay-based nanocomposites often obey continuum mechanics predictions. There are situations where nanocomposites can exhibit properties not expected with larger scale particulate reinforcements.

It is now well-recognized that the crystallization rate and degree of crystallinity can be influenced by crystallization in confined spaces. In these cases, the dimensions available for spherulitic growth are confined such that primary nuclei are not present for heterogeneous crystallization and homogeneous nucleation thus results. This results in the value of n in the Avrami equation approaching one and often leads to reduced crystallization rate, degree of crystallinity and melting point. This has been observed in phase separated block copolymers [7] and [8] and has also been observed in polymer blends [9]. Confined crystallization of linear polyethylene in nanoporous alumina showed homogeneous nucleation with pore diameters of 62–110 nm but heterogeneous nucleation for 15–48 nm pores [10]. Linear polyethylene [11] and syndiotactic polystyrene [12] in nanoporous alumina both showed decreased crystallinity versus bulk crystallization. With nanoparticle incorporation in a polymer matrix, similarities to confined crystallinity (as noted above for crystallization in nanopores) exist as well as nucleation effects and disruption of attainable spherulite size.

With inorganic particle and nanoparticle inclusions, nucleation of crystallization can occur. At the nanodimension scale, the nanoparticle can substitute for the absence of primary nuclei thus competing with the confined crystallization. At higher nanoparticle content, the increased viscosity (decreased chain diffusion rate) can lead to decreased crystallization kinetics. Thus, the crystallization process is complex and influenced by several competing factors. Nucleation of crystallization (at low levels of addition) evidenced by the onset temperature of crystallization (Tc) and crystallization half-time has been observed in various nanocomposites (poly(ε-caprolactone)–nanoclay [13], polyamide 66–nanoclay [14] and [15], polylactide-nanoclay [16], polyamide 6–nanoclay [17], polyamide 66–multi-walled carbon nanotube [18], polyester–nanoclay [19], poly(butylene terephthalate)–nanoclay [20], polypropylene–nanoclay (sepiolite) [21], polypropylene/multi-walled carbon nanotube [22]). At higher levels of nanoparticle addition, retardation of the crystallization rate has been observed even in those systems where nucleation was observed at low levels of nanoparticle incorporation [15], [18], [20], [22], [23] and [24]. The higher level of nanoparticle inclusion was noted to yield retardation of crystallization due to diffusion constraints. This was also apparent in a study where unmodified and organically modified clay were incorporated in maleic anhydride grafted polypropylene [25]. Nucleation was observed with unmodified clay, whereas the exfoliated clay yielded a reduced crystallization rate. A recent review of the crystallization behavior of layered silicate clay nanocomposites noted that while nucleation is observed in many systems the overall crystallization rate is generally reduced particularly at higher levels of nanoclay addition [26].

Another "nano-effect" noted in the literature has been the change in the Tg of the polymer matrix with the addition of nano-sized particles. Both increases and decreases in the Tg have been reported dependant upon the interaction between the matrix and the particle. In essence, if the addition of a particle to an amorphous polymer leads to a change in the Tg, the resultant effect on the composite properties would be considered a "nano-effect" and not predictable employing continuum mechanics relationships unless the Tg changes were properly accounted for or were quite minor. The glass transition of a polymer will be affected by its environment when the chain is within several nanometers of another phase. An extreme case of this is where the other environment is air (or vacuum). It has been well-recognized in the literature that the Tg of a polymer at the air–polymer surface or thin films (<100 nm) may be lower than that in bulk [27]. This can also be considered a confinement effect. A specific experimental example was reported where poly(2-vinyl pyridine) showed an increase in Tg, poly(methyl methacrylate) (PMMA) showed a decrease in Tg and polystyrene showed no change with silica nanosphere incorporation. These differences were ascribed to surface wetting [28]. The Tg decrease for PMMA was ascribed to free volume existing at the polymer surface interface due to poor wetting. In most literature examples where Tg values have been obtained, usually only modest changes are reported (<10 °C) as noted in various examples tabulated in Table 1. In some cases the organic modification of clay can result in a decrease in Tg due to plasticization [29]. It should be noted that the values noted in Table 1 involved relatively low levels of nanoparticle incorporation (<0.10 wt fraction and even lower volume fraction) and larger changes in Tg could be expected at much higher volume fraction loadings. For crosslinked polymers, another consideration is necessary as the presence of nanoparticles could yield a crosslink density change over the unmodified composite. This could be due to preferential interactions of the crosslinking agent with the nanoparticle surface or interruption of the crosslink density due to confinement effects. A theoretical model has been developed to predict the glass transition temperature of nanocomposites [30]. The model predicts both increases and decreases in Tg dependant upon specific interactions and shows good agreement with the experimental data noted above [28].

Table 1. Glass transition changes with nanofiller incorporation

Polymer	Nanofiller	T_g change (°C)	Reference
Polystyrene	SWCNT	3	[34]
Polycarbonate	SiC (0.5–1.5 wt%) (20–60 nm particles)	No change	[35]
Poly(vinyl chloride)	Exfoliated clay (MMT) (<10 wt%)	−1 to −3	[36]
Poly(dimethyl siloxane)	Silica (2–3 nm)	10	[37]
Poly(propylene carbonate)	Nanoclay (4 wt%)	13	[38]
Poly(methyl methacrylate)	Nanoclay (2.5–15.1 wt%)	4–13	[39]
Polyimide	MWCNT (0.25–6.98 wt%)	−4 to 8	[40]
Polystyrene	Nanoclay (5 wt%)	6.7	[41]
Natural rubber	Nanoclay (5 wt%)	3	[42]
Poly(butylene terephthalate)	Mica (3 wt%)	6	[43]
Polylactide	Nanoclay (3 wt%)	−1 to −4	[29]

SWCNT = single-walled carbon nanotubes; MMT = montmorillonite; MWCNT = multi-walled carbon nanotubes.

A situation does exist where significant increases in the glass transition temperature have been noted involving polyhedral oligomeric silsesquioxane (POSS) cage structures chemically reacted into the polymeric network [31], [32], [33] and [34]. These cage structures with a particle diameter in the range of 1–3 nm can be functionalized to provide chemical reactivity with various polymer systems. Examples include octavinyl (R = vinyl group) incorporation for copolymerization with PMMA [31], amine groups for incorporation into polyamides [32] and polyimides [33]. This parallels the glass transition increase often noted in the sol–gel inorganic–organic networks.

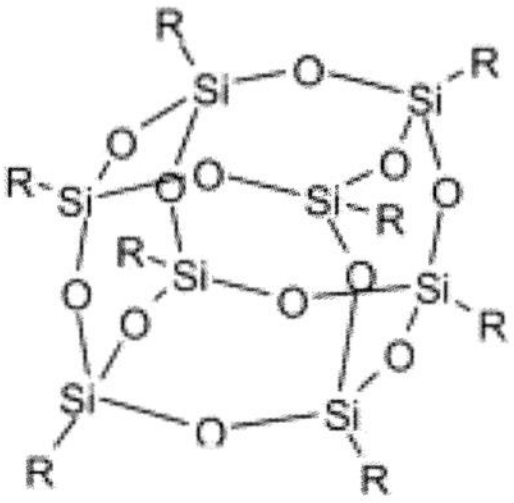

R = alkyl, aryl, cycloaliphatic, vinyl, amino, nitrile halogen. alcohol, ester, isocyanate, glycidyl etc.

POSS

While the glass transition temperature and crystallinity are the major property changes of interest of the nanocomposite polymer matrix, other "nano-effects" or property improvements over larger scale dimensions can be observed. Disruption of packing of rigid chain polymers resulting in higher free volume has been observed in permeability studies [44], surface area effects in photovoltaic applications involving conjugated polymers, surface area effects for catalysts incorporated in polymers, polymer chain dimensions where the radius of gyration is greater than the distance between adjacent nanoparticles, optical properties, nanofiber scaffolds for tissue engineering are additional areas. The "aging" of polymers is a thickness dependant property with rapid change at nanoscale dimensions [45] and [46]. This property is due to the ability of free volume to diffuse out of the sample and the diffusion coefficient (although very low) becomes important in the time scale associated with polymer utility (days to years) at nanoscale thicknesses. Surface area effects including catalysts, bioactivity, often require nanolevel dimensions to achieve optimum performance.

This review of polymer matrix based nanocomposites is divided into two major sections: clay-based nanocomposites with emphasis on mechanical reinforcement and other property modifications. Mechanical enhancement is usually associated with polymer-based composites, however, a number of other areas have emerged where additional property enhancements can be realized by incorporation of nanoscale particles, platelets or fibers.

CLAY-BASED POLYMER NANOCOMPOSITES

Structure of Montmorillonite

The clay known as montmorillonite consists of platelets with an inner octahedral layer sandwiched between two silicate tetrahedral layers [47] as illustrated in Fig. 1. The octahedral layer may be thought of as an aluminum oxide sheet where some of the aluminum atoms have been replaced with magnesium; the difference in valences of Al and Mg creates negative charges distributed within the plane of the platelets that are balanced by positive counterions, typically sodium ions, located between the platelets or in the galleries as shown in Fig. 1. In its natural state, this clay exists as stacks of many platelets. Hydration of the sodium ions causes the galleries to expand and the clay to swell; indeed, these platelets can be fully dispersed in water. The sodium ions can be exchanged with organic cations, such as those from an ammonium salt, to form an organoclay [48], [49], [50], [51], [52], [53], [54], [55], [56] and [57]. The ammonium cation may have hydrocarbon tails and other groups attached and is referred to as a "surfactant" owing to its amphiphilic nature. The extent of the negative charge of the clay is characterized by the cation exchange capacity, i.e., CEC.

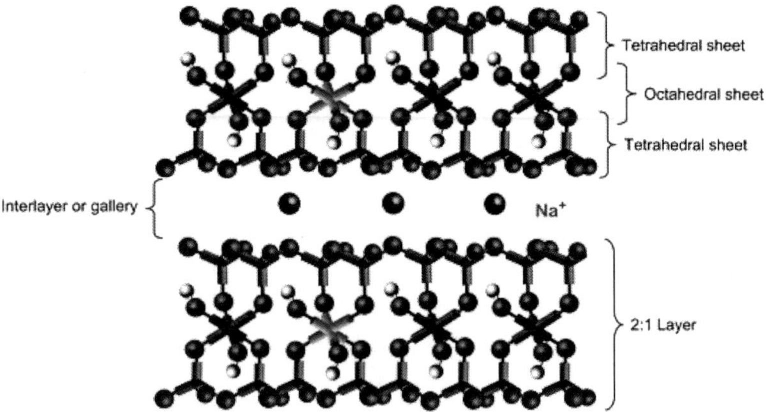

Figure 1. Structure of sodium montmorillonite. Courtesy of Southern Clay Products, Inc.

The X-ray d -spacing of completely dry sodium montmorillonite is 0.96 nm while the platelet itself is about 0.94 nm thick [47] and [58]. When the sodium is replaced with much larger organic surfactants, the gallery expands and the X-ray d -spacing may increase by as much as 2 to 3-fold [59] and [60]. While the thickness of montmorillonite platelets is a well-defined crystallographic dimension, the lateral dimensions of the platelets are not. They depend on how the platelets grew from solution in the geological process that formed them. Many authors grossly exaggerate the lateral size with dimensions quoted of the order of microns or even tens of microns. A commonly used montmorillonite was accurately characterized recently by depositing platelets on a mica surface from a very dilute suspension and then measuring the lateral dimensions by atomic force microscopy [61]. Since the platelets are not uniform or regular in lateral size or shape, the platelet area, A , was measured and its square-root was normalized by platelet thickness, t , to calculate an "aspect ratio". The distribution of aspect ratios found is shown in Fig. 2. If each platelet were circular with diameter D , then $\sqrt{A}/t$ source would be $\sqrt{\pi/4}(D/t) = 0.89(D/t)$ .. Since t is approximately 1 nm, Fig. 2 shows that the most probable lateral dimension is in the range of 100–200 nm.

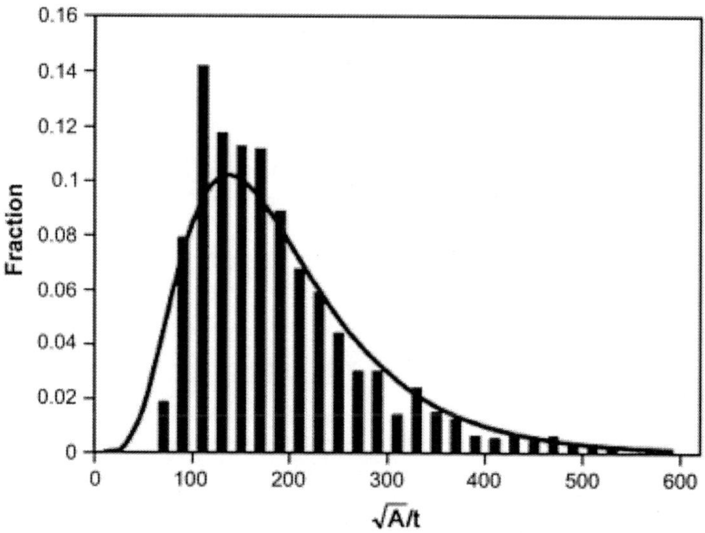

Figure 2. Aspect ratio distribution of native sodium montmorillonite platelets [61]. Reproduced with permission of the American Chemical Society.

Nanocomposite formation: exfoliation

Nanocomposites can, in principle, be formed from clays and organoclays in a number of ways including various in situ polymerization [4], [6], [62], [63], [64], [65], [66], [67] and [68], solution [51] and [53], and latex [69] and [70] methods. However, the greatest interest has involved melt processing [71], [72], [73], [74], [75], [76], [77], [78], [79], [80], [81], [82], [83], [84], [85], [86], [87], [88], [89], [90], [91], [92], [93], [94], [95], [96], [97], [98], [99], [100], [101], [102], [103], [104], [105], [106], [107], [108], [109], [110], [111], [112], [113], [114], [115], [116], [117], [118], [119], [120], [121], [122], [123], [124], [125], [126], [127], [128], [129], [130], [131], [132], [133], [134], [135], [136], [137], [138] and [139] because this is generally considered more economical, more flexible for formulation, and involves compounding and fabrication facilities commonly used in commercial practice. For most purposes, complete exfoliation of the clay platelets, i.e., separation of platelets from one another and dispersed individually in the polymer matrix, is the desired goal of the formation process. However, this ideal morphology is frequently not achieved and varying degrees of dispersion are more common. While far from a completely accurate or descriptive nomenclature, the literature commonly refers to three types of morphology: immiscible (conventional or microcomposite), intercalated, and miscible or exfoliated. These are illustrated schematically in Fig. 3 along with example transmission electron microscopic, TEM, images and the expected wide angle X-ray scans [48], [49], [50], [51], [52], [53] and [83].

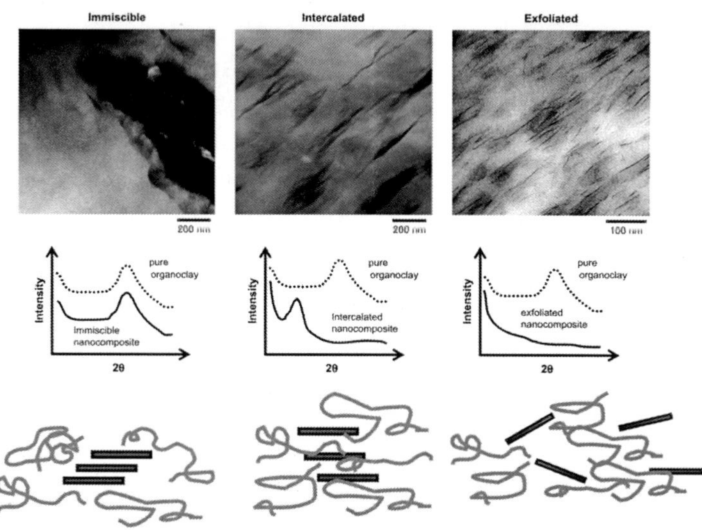

Figure 3. Illustration of different states of dispersion of organoclays in polymers with corresponding WAXS and TEM results.

For the case called "immiscible" in Fig. 3, the organoclay platelets exist in particles comprised of tactoids or aggregates of tactoids more or less as they were in the organoclay powder, i.e., no separation of platelets. Thus, the wide angle X-ray scan of the polymer composite is expected to look essentially the same as that obtained for the organoclay powder; there is no shifting of the X-ray d-spacing. Generally, such scans are made over a low range of angles, 2θ, such that any peaks from a crystalline polymer matrix are not seen since they occur at higher angles. For completely exfoliated organoclay, no wide angle X-ray peak is expected for the nanocomposite since there is no regular spacing of the platelets and the distances between platelets would, in any case, be larger than what wide angle X-ray scattering can detect.

Often X-ray scans of polymer nanocomposites show a peak reminiscent of the organoclay peak but shifted to lower 2θ or larger d-spacing. The fact that there is a peak indicates that the platelets are not exfoliated. The peak shift indicates that the gallery has expanded, and it is usually assumed that polymer chains have entered or have been intercalated in the gallery. Placing polymer chains in such a confined space would involve a significant entropy penalty that presumably must be driven by an energetic attraction between the polymer and the organoclay [76], [77], [78] and [79]. It is possible that the gallery expansion may in some cases be caused by intercalation of oligomers or low molecular weight polymer chains. The early literature seemed to suggest that "intercalation" would be useful and perhaps a precursor to exfoliation. Subsequent research has suggested alternative ideas about how the exfoliation process may occur in melt processing and how the details of the mixing equipment and conditions alter the state of dispersion achieved [54], [82], [84] and [140]. These ideas are summarized in the cartoon shown in Fig. 4 [84]. As made commercially, the particles of an organoclay powder are about 8 μm in size and consist of aggregates of tactoids, or stacks of platelets; the stresses imposed during melt mixing break up aggregates and can shear the stack into smaller ones as suggested in Fig. 4. However, there evidently is a limit to how finely the clay can be dispersed just by mechanical forces. If the polymers and organoclay have an "affinity" for one another, the contact between polymers and organoclay can be increased by peeling the platelets from these stacks one by one until, given enough time in the mixing device, all the platelets are individually dispersed as suggested in Fig. 4. This notion is supported by many TEM images at various locations in the extruder and is more plausible than imagining the polymer chains diffusing into the galleries, i.e., intercalation, and eventually pushing them further and further apart until an exfoliated state is reached.

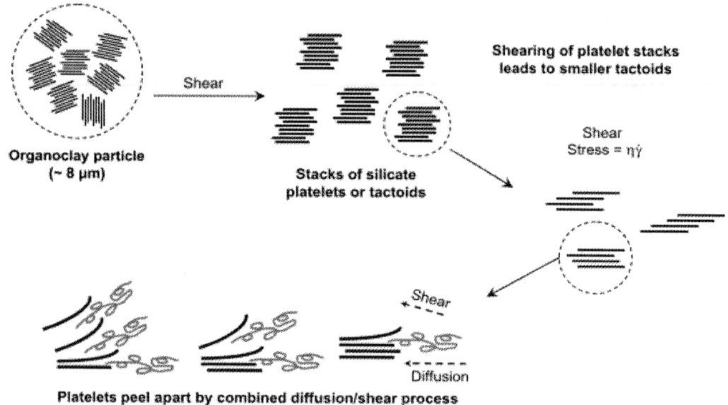

Figure 4. Mechanism of organoclay dispersion and exfoliation during melt processing [84]. Reproduced with permission of Elsevier Ltd.

The nature of the extruder and the screw configuration are important to achieve good organoclay dispersion [83]. Longer residence times in the extruder favor better dispersion [83]. In some cases, having a higher melt viscosity is helpful in achieving dispersion apparently because of the higher stresses that can be imposed on the clay particles [84] and [126]; however, this effect is not universally observed. The location of where the organoclay is introduced into the extruder has also been shown to be important [120]. However, no matter how well these process considerations are optimized, it is clear that complete exfoliation, or nearly so, cannot be achieved unless there is a good thermodynamic affinity between the organoclay and the polymer matrix. This affinity can be affected to a very significant extent by optimizing the structure of the surfactant used to form the organoclay [85], [88], [99], [100], [109], [113], [119] and [141] and possibly certain features of the clay itself like its CEC [115], as this affects the density of surfactant molecules over the silicate surface.

A key factor in the polymer–organoclay interaction is the affinity polymer segments have for the silicate surface [84], [85], [94], [113], [141] and [142]. Nylon 6 appears to have good affinity for the silicate surface, perhaps by hydrogen bonding, and as a result very high levels of exfoliation can be achieved in this matrix provided the processing conditions and melt rheology are properly selected [83], [84] and [120]. Surfactants with a single long alkyl tail give the best exfoliation [141]. As more long chain alkyls are added to the surfactant, the extent of exfoliation is decreased [141]. It has been proposed that at least one alkyl tail is needed to reduce the platelet-platelet cohesion while adding more than one tends to block access of the polyamide chains from the silicate surface diminishing these favorable interactions while increasing the very unfavorable alkyl-polyamide interaction. On the other hand, non-polar polyolefin segments have no attraction to the polar silicate surface, and in this case, increasing the number of alkyls on the surfactant improves dispersion of the organoclay in the polyolefin matrix since a larger number of alkyls decrease the possible frequency of the unfavorable polyolefin–silicate interaction and increases the frequency of more favorable polyolefin–alkyl contacts [96], [100] and [105].

Even under the best of circumstances exfoliation of organoclays in neat polyolefins like polypropylene, PP, or polyethylene, PE, is not very good and far less than that observed in polyamides, polyurethanes, and some other polar polymers [84], [96] and [101]. It has been found that a small amount of a polyolefin that has been lightly grafted with maleic anhydride, ~1% MA by weight is typical, can act as a very effective "compatibilizer" for dispersing the organoclay in the parent polyolefin [84], [101], [103], [106], [117], [118], [143], [144], [145], [146], [147] and [148]. This does not lead to the high level of exfoliation that can be achieved in polyamides, but this approach has allowed such nanocomposites to move forward in commercial applications, particularly in automotive parts [54], [99] and [106].

In the case of olefin copolymers with polar monomers like vinyl acetate and methacrylic acid (and corresponding ionomers), the degree of exfoliation that can be achieved progressively improves as the polar monomer content increases [113] and [119]. In all cases, the best exfoliation is achieved when the structure of the surfactant and the process parameters are optimized.

Characterization of nanocomposite morphology

An important issue is to relate the performance of nanocomposites to their morphological structure; experimental evaluation of performance is certainly easier than characterization of their morphology. Wide angle X-ray scattering, WAXS, is frequently used because such analyses are relatively simple to do. However, such analyses can be misleading and are not quantitative [149], [150] and [151]. As indicated in Fig. 3, the organoclay has a characteristic peak indicative of the platelet separation or d-spacing; other peaks may be seen resulting from multiple reflections as predicted by Bragg's law. The presence of the same peak in the nanocomposite is irrefutable evidence that the nanocomposite contains organoclay tactoids as suggested in Fig. 3. However, the absence of such a peak is not conclusive evidence for a highly exfoliated structure as has been repeatedly pointed out in the literature [151]; many factors must be considered to interpret WAXS scans. If the sensitivity, or counting time, of the scan is low, then an existing peak may not be seen. When the tactoids are internally disordered or not well aligned to one another, the peak intensity will be low and may appear to be completely absent. These issues can be well illustrated by analyses of polyolefin nanocomposites, which are never fully exfoliated, that have been injection molded. X-ray scans of the molded surface reveal a peak indicating the presence of tactoids. However, after milling away the surface of these specimens, subsequent scans of the milled surface in the core of the bar may not reveal a peak because the tactoids are more randomly oriented in the interior than near the as-molded surface [103] and [118]. However, if a more sensitive scan is made, the peak can usually be seen.

In some cases, the WAXS scan may reveal a shift in the peak location relative to that of the neat organoclay. The peak may shift to lower angles, or larger d-spacing, and is generally taken as evidence of "intercalation" of polymers (or perhaps other species) into the galleries [48], [49], [50], [51], [52], [53], [73] and [79]. However, an opposite shift may also occur, and this is usually attributed to loss of unbound surfactant from the gallery or to surfactant degradation [89] and [107]. All of these processes may occur simultaneously rendering uncertainty in the interpretation. In any case, intercalation per se does not seem to be a contributor to develop useful nanocomposite performance.

Small angle X-ray scattering, SAXS, can be more informative and somewhat quantitative as explained by numerous authors [17], [152], [153], [154], [155] and [156]. However, this technique has not been widely used except in a few laboratories probably because most laboratories do not have SAXS facilities or experience in interpreting the results. Other techniques like solid-state NMR and neutron scattering have also been used on a limited basis to explore clay dispersion [95], [157], [158], [159], [160], [161] and [162].

A far more direct way of visualizing nanocomposite morphology is via transmission electron microscopy, TEM; however, this approach requires considerable skill and patience but can be quantitative. Use of TEM is often criticized because it reveals the morphology in such a small region. However, this can be overcome by taking images at different magnifications and from different locations and orientations until a representative picture of the morphology is established. The major obstacle in obtaining good TEM images is not in the operation of the microscope but in microtoming sections that are thin and uniform enough to reveal the morphology. Fortunately, the elemental composition of the clay compared to that of the polymer matrix is such that no staining is required. When exfoliation is essentially complete, as in the case of nylon 6, one can see the ~1 nm thick clay platelets as dark lines when the microtome cut is perpendicular to the platelets. Image analysis can be used to quantify the distribution of platelet lengths, but meaningful statistics require analyzing several hundred particles [58], [84], [93] and [119]. However, it must be remembered that the dimensions observed reflect a random cut through an irregular platelet and only rarely will the maximum dimension be seen [163] and [164]. Thus, the aspect ratio distribution seen in this way will lead to smaller values than true dimensions like those given by Fig. 2.

Even for the best nylon 6 nanocomposites, exfoliation is generally never complete and one can see particles consisting of two, three or more platelets [58]. In some cases, these platelets may be skewed relative to one another as suggested in Fig. 5[93]. Thus, some particles may appear to be longer than the platelets really are. These kinds of issues should be kept in mind when interpreting quantitative analyses of particle aspect ratios and in comparison of observed performance with that predicted by composite theory [58].

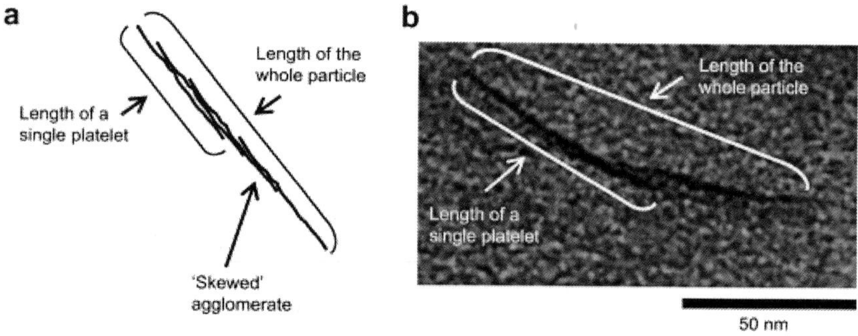

Figure 5. Examples of skewed platelets such that particles appear longer than platelets of MMT [93]. Reproduced with permission of Elsevier Ltd.

Nanocomposites made from polyolefins, styrenics, and other polymers that lead to lower degrees of exfoliation reveal particles much thicker than single clay platelets as expected [101], [110], [111], [117] and [119]. However, the clay particles are also much longer than the individual clay platelets indicated in Fig. 2. As the polymer-organoclay affinity is increased by adding a compatibilizer, e.g., PP-g-MA or PE-g-MA, or increasing the content of a polar comonomer, e.g., vinyl acetate, the clay particles not only become thinner (fewer platelets in the stack) but also become shorter [103], [106], [117], [118], [119] and [120]. However, generally the particle thickness decreases more rapidly than the length such that the aspect ratio increases; this generally improves performance.

The fact that the particles become shorter does not mean that clay platelets are breaking or being attributed during processing, although, this may occur under some extreme conditions [120]. Instead, considerable evidence indicates that the vision of tactoids as usually drawn, see Fig. 3 and Fig. 4, where the platelets are all of the same length and in registry with one another is not correct. Fig. 6 shows a more realistic vision of a tactoid where the particle length can be much longer than individual platelets and how these particles evolve as dispersion improves [106] and [117].

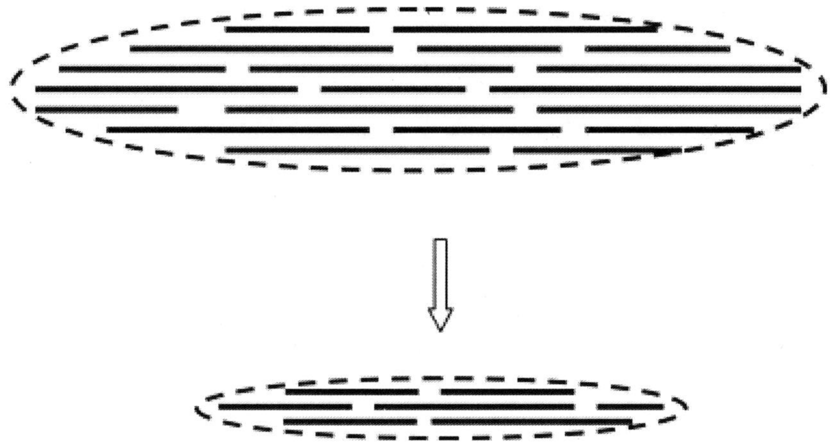

Figure 6. A more realistic picture of clay tactoids and how they become shorter as the level of dispersion increases.

Complications arise when calculating an average aspect ratio of particles when there is a distribution of both length and thickness. First, one can calculate a number average, a weight average, or other weightings of the distribution [58], [113] and [119]. Second, one can average the aspect ratios or average separately the lengths and thickness and calculate an aspect ratio from these averages [113] and [119]. There is no theoretical guidance on which is the better predictor of performance or for use in composite modeling [113], [117] and [119].

To take full advantage of the reinforcement or tortuosity clay platelets or particles can provide to mechanical and thermal or barrier properties of nanocomposites, they must be oriented in the appropriate direction and not curled or curved. The alignment of particles is affected by the type of processing used to form the test specimen, e.g., extrusion, injection molding, etc. This is a separate issue from the degree of dispersion or exfoliation which is usually determined in the mixing process. Techniques like compression molding usually do not lead to good alignment or straightening of the high aspect ratio particles, and measurements made on such specimens often underestimate the potential performance. TEM can be used to assess and even quantify particle orientation and curvature and this information can, in principle, be factored into appropriate models to ascertain their effect on performance [86], [110], [131], [134] and [138].

Nanocomposite mechanical properties: reinforcement

A common reason for adding fillers to polymers is to increase the modulus or stiffness via reinforcement mechanisms described by theories for composites [58], [165], [166], [167], [168], [169], [170], [171], [172], [173], [174], [175], [176], [177], [178], [179], [180], [181], [182], [183], [184] and [185]. Properly dispersed and aligned clay platelets have proven to be very effective for increasing stiffness. This is illustrated in Fig. 7 by comparing the increase in the tensile modulus, E, of injection molded composites based on nylon 6, relative to the modulus of the neat polyamide matrix, Em, when the filler is an organoclay versus glass fibers [58]. In this example, increasing the modulus by a factor of two relative to that of neat nylon 6 requires approximately three times more mass of glass fibers than that of montmorillonite, MMT, platelets. Thus, the nanocomposite has a weight advantage over the conventional glass fiber composite. Furthermore, if the platelets are aligned in the plane of the sample, the same reinforcement should be seen in all directions within the plane, whereas fibers reinforce only along a single axis in the direction of their alignment [165]. In addition, the surface finish of the nanocomposite is much better than that of the glass fiber composite owing to nanometer size of the clay platelets versus the 10–15 μ diameter of the glass fibers. A central question is whether the greater efficiency of the clay has anything to do with its nanometric dimensions, i.e., a "nano-effect". To answer this requires considering many issues which we will do later in this section; however, the short answer is that we can explain essentially all of the experimental trends using composite theory without invoking any "nano-effects" [58].

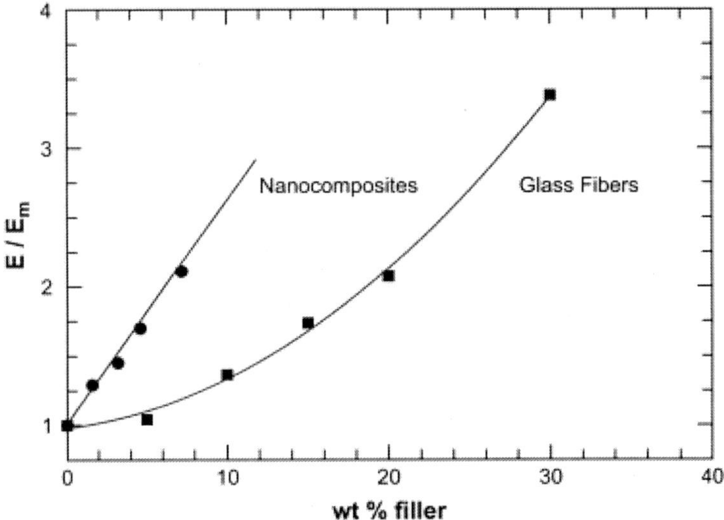

Figure 7. Comparison of modulus reinforcement (relative to matrix polymer) increases for nanocomposites based on MMT versus glass fiber (aspect ratio ~20) for a nylon 6 matrix [58]. Reproduced by permission of Elsevier Ltd.

Fig. 8 shows an analogous comparison of nanocomposites based on thermoplastic polyolefin or TPO matrix, polypropylene plus an ethylene-based elastomer, with conventional talc-filled TPO [103]. The latter is widely used in automotive applications; however, in some cases, they are being replaced with TPO nanocomposites. In this case, doubling the modulus of the TPO requires more than four times more talc than MMT; this presents a weight, and consequently fuel, savings along with the improved surface finish [103] and [106]. The polyamide nanocomposites in Fig. 7 are very highly exfoliated, whereas the exfoliation of the clay in the TPO of Fig. 8 is not nearly so perfect [103], [106], [117] and [118]. However, it should be recognized that the talc particles do not have as high aspect ratio as the glass fibers used in these comparisons [58] and [186]. Another factor at play here is the lower modulus of TPO than nylon 6. The lower the matrix modulus, the greater is the relative increase in reinforcement caused by adding a filler [94].

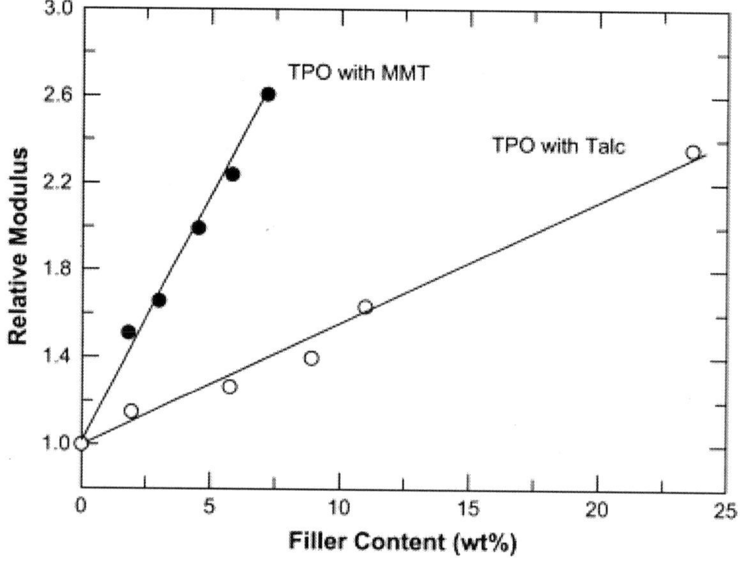

Figure 8. Comparison of modulus reinforcement for nanocomposites based on MMT versus talc for a TPO matrix [103]. Reproduced by permission of Elsevier Ltd.

Fig. 9 shows dynamic mechanical moduli of the nylon 6 nanocomposites from Fig. 7 versus temperature. The intersection of these curves with the horizontal line shown is a good approximation to the heat distortion temperature, HDT, of these materials [58]. This temperature, used as a benchmark for many applications, can be increased by approximately 100 °C by addition of about 7% by weight of MMT. This effect has been explained by simple reinforcement, as predicted by composite theory, without invoking any special "nano-effects"; the effect of MMT on the glass transition of these materials is very slight if any at all [58]. Indeed, glass fibers cause an analogous increase in HDT.

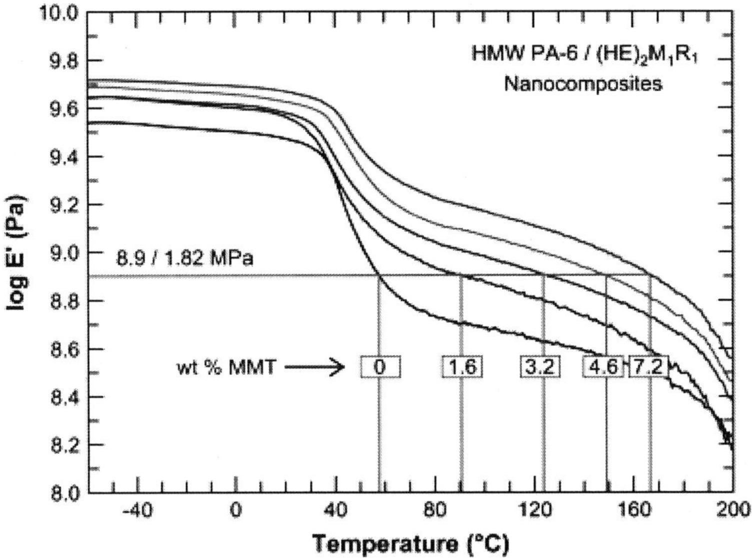

Figure 9. Experimental storage modulus data versus temperature for nylon 6 nanocomposites. The horizontal line is used to estimate the heat distortion temperature (HDT) at an applied stress of 1.82 MPa or 264 psi [58]. Reproduced by permission of Elsevier Ltd.

Addition of fillers, including clay, can also increase strength as well as modulus [84]; however, the opposite may also occur [99]. A main issue is the level of adhesion of the filler to the matrix. For glass fiber composites, chemical bonding at the interface using silane chemistry is used to achieve high strength composites [186]. On the other hand, the modulus of glass fiber composites is not very much affected by the level of interfacial adhesion [186]. Unfortunately, at this time there is no effective way to measure the level of adhesion of clay particles with polymer matrices. In addition, there are no effective methods at this time to create chemical bonds between clay particles and polymer matrices analogous to those used for glass fibers. Generally, addition of organoclays to ductile polymers increases the yield strength; however, for brittle matrices failure strength is typically decreased [84], [99], [100] and [101].

Addition of fillers generally decreases the ductility of polymers, e.g., elongation at break. For glass fibers, talc, etc. this is well known and expected. Similar trends are also seen for nanocomposites [84] and

[100], but this seems to have been unexpected and disappointing to some working in this field. Impact strength is an energy measurement, i.e., a force acting through a distance. A reduced elongation at break often means a reduced energy to break but there are exceptions to this [84], [100] and [116]. Addition of clay may increase the stress levels via reinforcement more than the reduction in deformation as recently demonstrated for some nanocomposites [116]. Generally speaking, the reduction in ductility or energy to break is more severe when the polymer matrix is below its glass transition, whereas the effects of adding clay may not be so dramatic when the matrix is above its glass transition temperature [82], [84], [100] and [119]. This involves a shift in fracture mechanisms that is beyond the scope of this review.

Melt rheological properties of polymers can be dramatically altered in the low shear rate or frequency region such that these fluids appear to have a yield stress [84], [117], [118], [187], [188] and [189]. The effects in the high shear rate region are usually much less dramatic [84]. A great deal has been written about these effects and their causes, and this will not be reviewed here. Interestingly, the addition of clay seems to be an effective way to increase "melt strength" which can be useful in some polymer processing operations like film blowing or blow molding [54] and [96].

To answer the question of whether the large increase in modulus caused by clay platelets or particles relative to conventional fillers, like that illustrated in Fig. 7 and Fig. 8, is due to some "nano-effect", one must first determine whether effects of this magnitude can be predicted by composite theories. That is, by a "nano-effect", we mean some change in the local properties of the matrix caused by the extremely high surface area filler and the small distances between nanofiller particles even at low mass loadings. It is well known that clay particles are effective nucleating agents which greatly change the crystalline morphology and crystal type for polymers like nylon 6 or PP [87]. Potential "confinement" effects are also discussed in the context of nanocomposites.

A basic premise of composite theories is that the matrix and filler have the same properties as when the other component is not there. These theories, thus, only predict the effects of simple reinforcement and do not allow for any "nano-effects" of the type mentioned.

Clearly, reinforcement does occur and the issue is whether that alone can explain the observations or not. Composite theories consider only the aspect ratio, orientation and volume fraction of filler in the matrix; the absolute filler particle size does not enter into the calculations. We have already mentioned the difficulties of experimentally determining the aspect ratio (sectioning issues, averaging of distributions, etc.). Furthermore, determining what values to assign to the properties of the clay platelets (like its modulus) is not trivial. Finally, the various composite theories differ somewhat in their predictions owing to the assumptions and simplifications used in their mathematical formulation. These and other issues are worth remembering as we proceed with their analysis using data for nylon 6 nanocomposites.

We wish to compare composite calculations with experimental data for the modulus of nylon 6 nanocomposites where the degree of exfoliation is very high but not perfect. An image analysis of many TEM photomicrographs was used to construct a platelet length distribution which looks very similar to that in Fig. 2[58]; the number average platelet length was found to be 91 nm. While the majority of the clay particles was single platelets (thickness ~ 0.94 nm), there were some doublets, some triplets, and a few quadruplets. It was estimated, by a rather involved analysis, that the number average platelet thickness was 1.61 nm [58]. Thus, an upper bound on the aspect ratio for perfect exfoliation (using number averages of the distribution) would be about 91/0.94 = 97 while a more realistic estimate might be 91/1.61 = 57.

Next, we need the in-plane modulus of a montmorillonite platelet. Information from a variety of sources suggests that a reasonable value is 178 GPa [58] and [190]; however, some molecular dynamics calculations suggest significantly larger values [191]. A density for MMT of 2.83 g/cm3 was used to convert weight fractions to volume fractions. The properties of the matrix (modulus, Poisson ratio, density, etc.) were experimentally measured values [58]. The equations of Halpin–Tsai [168] and Mori–Tanaka [167] are frequently used for composite calculations; the former predicts higher levels of reinforcement for the cases of interest here than the latter as seen in Fig. 10. Interestingly, the predictions via these two theories and the two estimates of aspect ratio give results that bracket the experimental data. Thus, we conclude that simple reinforcement considerations adequately explain the observations given all the issues involved in making these calculations. Any "nano-effect" is relatively minor if at all.

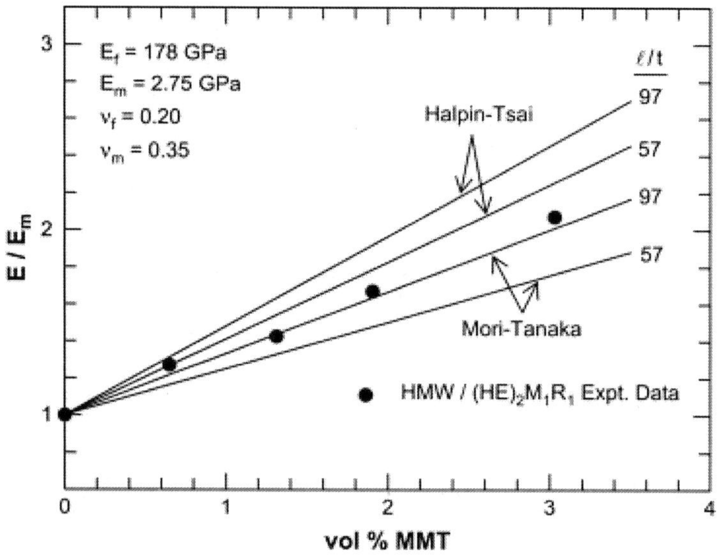

Figure 10. Experimental and theoretical stiffness data for nylon 6 nanocomposites; model predictions are based on unidirectional reinforcement of pure MMT having a filler modulus of 178 GPa and aspect ratio of 57 (experimentally determined number average value) and 97, corresponding to complete exfoliation. Note that experimental modulus data are plotted versus vol% MMT since MMT is the reinforcing agent [58]. Reproduced by permission of Elsevier Ltd.

More needs to be said about the comparisons shown in Fig. 7 and Fig. 8. The aspect ratio of the glass fibers in the nylon 6 matrix is about 20, whereas the aspect ratio of MMT platelets is 3–5 times larger than this. However, calculations using composite theory reveal that the larger aspect ratio of the clay versus glass fibers is not enough to explain all of the large differences in modulus reinforcement shown in Fig. 7. A significant part of the difference in modulus enhancement stems from the much higher modulus of MMT than glass fibers, i.e., 178 versus 72.4 GPa [58]. The comparison of MMT versus talc in a TPO matrix shown in Fig. 8 is more complex to explain, but similar factors are at play [103] and [117].

Nanocomposite thermal properties: dimensional stability

The high thermal expansion coefficients of neat plastics causes dimensional changes during molding and as the ambient temperature changes that are either undesirable or in some cases unacceptable for certain applications. The latter is a particular concern for automotive parts where plastics must be integrated with metals which have much lower coefficients of thermal expansion, CTE. Fillers are frequently added to plastics to reduce the CTE. For low aspect ratio filler particles, the reduction in CTE follows, more or less, a simple additive rule and is not very large; in these cases, the linear CTE changes are similar in all three coordinate directions. However, when high aspect ratio fillers, like fibers or platelets, are added and well oriented, the effects can be much larger; in these cases, the CTE in the three coordinate directions may be very different.

The fibers or platelets typically have a higher modulus and a lower CTE than the matrix polymer. As the temperature of the composite changes, the matrix tries to extend or contract in its usual way; however, the fibers or platelets resist this change creating opposing stresses in the two phases. When the filler to matrix modulus is large, the restraint to dimensional change can be quite significant within the direction of alignment. Platelets can provide their restraint in two directions, when appropriately oriented, while fibers can only do so in one direction. Because of their shape differences, fibers can cause a greater reduction in the direction of their orientation than platelets can [166] and [170]. The CTE in the direction normal to the fibers or the platelet plane can actually increase when such fillers are added. Theories based on the mechanisms described above are available for quantitatively predicting CTE behavior [166] and [170].

Montmorillonite platelets are particularly effective for reducing CTE of plastics as shown in Fig. 11 for well-exfoliated nylon 6 nanocomposites [86]. These data were measured in the flow direction of injection molded bars. When the semicrystalline nylon 6 matrix is above its glass transition temperature, the CTE reduction is greater than when below the Tg. Of course, the neat nylon 6 has a higher CTE above Tg than below; however, because of its lower modulus above Tg, the MMT platelets are more effective for reducing CTE. Note that the two curves in Fig. 11 seem to cross at about 7 wt% MMT. For these specimens, the CTE in the transverse direction is also reduced by adding MMT but not quite as efficiently as in the flow direction since platelet orientation is not as great in the former as the latter direction. The CTE in the normal direction actually increases as MMT is added. These trends are quantitatively predicted by the theories mentioned earlier [86].

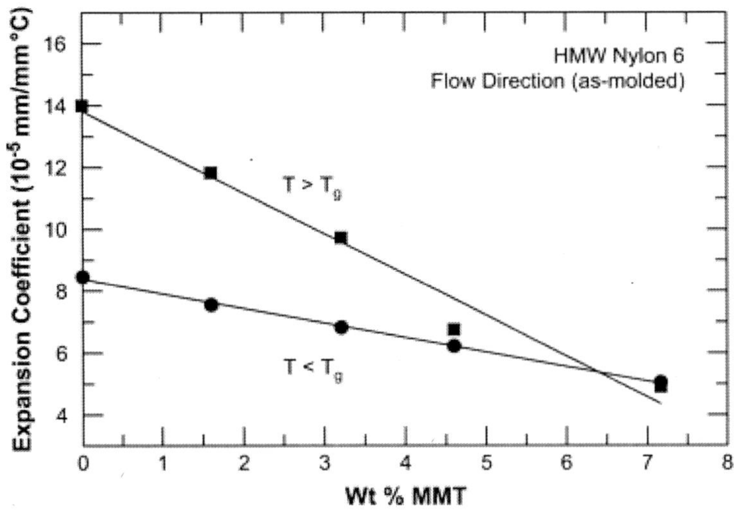

Figure 11. Linear thermal expansion coefficients of as-molded nylon 6 nanocomposites determined in the flow direction for T > Tg and T < Tg [86]. Reproduced by permission of Elsevier Ltd.

CTE behavior is also a major consideration for the TPO materials used in automotive applications [103], [106] and [117]. As seen in Fig. 12, MMT is much more efficient at reducing CTE than talc in these materials [106]. Again, composite theories capture these trends reasonably well [117].

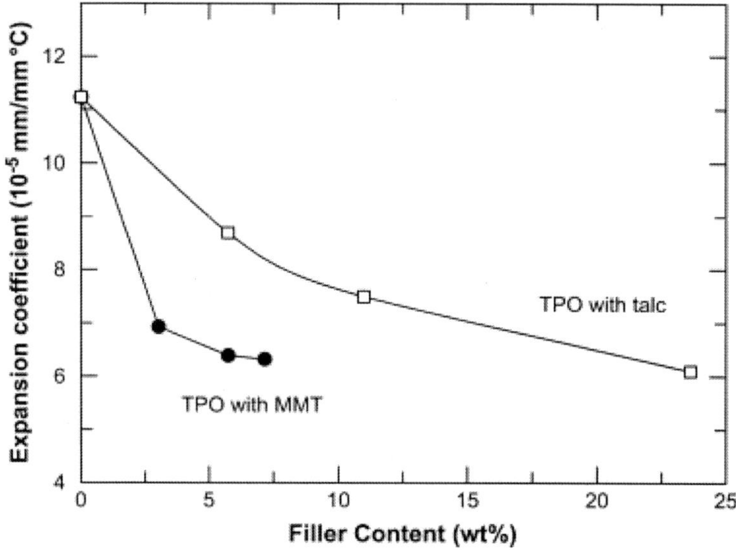

Figure 12. Comparison of linear coefficients of thermal expansion (flow directions) as a function of filler content of TPO composites formed from MMT and talc [106]. Reproduced by permission of Elsevier Ltd.

VARIATIONS AND APPLICATIONS OF POLYMER-BASED NANOCOMPOSITES: PROPERTIES OTHER THAN REINFORCEMENT

Polymer composites comprising nanoparticles (including nanofibers where the fiber diameter is in the nanodimension range) are often investigated where reinforcement of the polymer matrix is achieved. While the reinforcement aspects are a major part of the nanocomposite investigations reported in the literature, many other variants and property enhancements are under active study and in some cases commercialization. The advantages of nanoscale particle incorporation can lead to a myriad of application possibilities where the analogous larger scale particle incorporation would not yield the sufficient property profile for utilization. These areas include barrier properties, membrane separation, UV screens, flammability resistance, polymer blend compatibilization, electrical conductivity, impact modification, and biomedical applications. Examples of nanoparticle, nanoplatelet and nanofiber incorporation into polymer matrices are listed in Table 2 along with potential utility where properties other than mechanical property reinforcement are relevant.

Table 2. Examples of nanoscale filler incorporated in polymer composites for property enhancement other than reinforcement

Nanofiller	Property enhancement(s)	Application/utility
Exfoliated clay	Flame resistance, barrier, compatibilizer for polymer blends	
SWCNT; MWCNT	Electrical conductivity, charge transport,	Electrical/electronics/ optoelectronics
Nanosilver	Antimicrobial	
ZnO	UV adsorption	UV screens
Silica	Viscosity modification	Paint, adhesives
CdSe, CdTe	Charge transport	Photovoltaic cells
Graphene	Electrical conductivity, barrier, charge transport	Electrical/electronic
POSS	Improved stability, flammability resistance	Sensors, LEDs

Barrier and Membrane Separation Properties

The barrier properties of polymers can be significantly altered by inclusion of inorganic platelets with sufficient aspect ratio to alter the diffusion path of penetrant molecules as illustrated in Fig. 13. Various continuum models have been proposed to predict the permeability of platelet filled composites as listed in Table 3. These models are generally based on random, parallel platelets perpendicular to the permeation direction (random in only two directions). The model by Bharadwaj introduces an orientation factor [196]. At high aspect ratio which can be achieved (such as with exfoliated clay) in nanocomposites, significant decreases in permeability are predicted and observed in practice. Four of these models have been applied to polyisobutylene/vermiculite nanocomposites with aspect ratios predicted in the range of expectations [70]. The variability in the models was, however, shown to be substantial. In many cases, the nanocomposites investigated can be approximated by the continuum models, thus the "nano-effect" is not observed. This should not be surprising as the dimensions of permeating gas molecules are still much lower than the nanodimension modification. Differences would be expected in those cases where the Tg of the matrix polymer is changed. However, for practical applications the nanoscale dimensions are still quite important as transparency can be maintained along with surface smoothness for thin films; critical for food packaging applications.

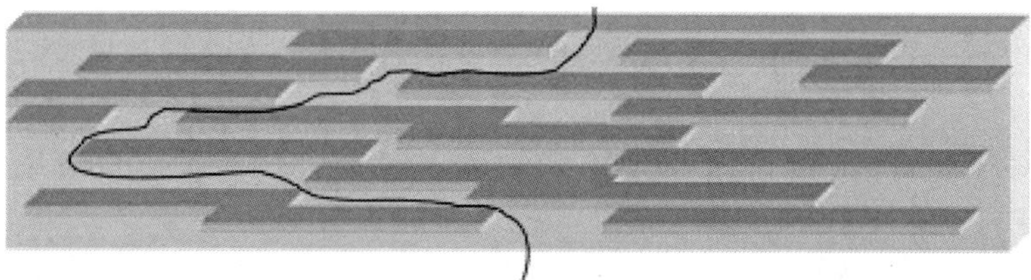

Permeation path imposed by nanoplatelet modification of polymer films

Figure 13. Barrier to permeation imposed by nanoparticles imbedded in a polymeric matrix.

Table 3. Models for predicting barrier properties of platelet filled nanocomposites (adapted from Ref. [70], copyright by Elsevier)

Model	Filler type	Particle geometry	Formulas	Reference
Nielsen	Ribbon[a]	w t	$(P_0/P)(1-\phi) = 1 + \alpha\phi/2$	[192]
Cussler				
(Regular array)	Ribbon[a]	w t	$(P_0/P)(1-\phi) = 1 + (\alpha\phi)^2/4$	[193]
(Random array)	Ribbon[a]	w t	$(P_0/P)(1-\phi) = (1 + \alpha\phi/3)^2$	[193]
Gusev and Lusti	Disk[b]	d t	$(P_0/P)(1-\phi) = \exp[(\alpha\phi/3.47)^{0.71}]$	[194]

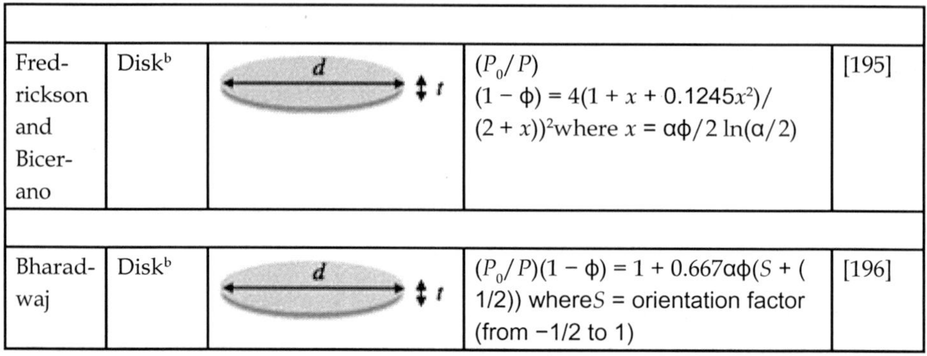

Fredrickson and Bicerano	Disk[b]	d, t	(P_0/P) $(1-\phi) = 4(1 + x + 0.1245x^2)/$ $(2 + x))^2$where $x = \alpha\phi/2\ \ln(\alpha/2)$	[195]
Bharadwaj	Disk[b]	d, t	$(P_0/P)(1-\phi) = 1 + 0.667\alpha\phi(S + ($ $1/2))$ where$S$ = orientation factor (from −1/2 to 1)	[196]

a For ribbons, length is infinite, width, w; thickness, t; aspect ratio, $\alpha \equiv w/t$.

b For disks, circular shape of diameter d and thickness t; aspect ratio, $\alpha \equiv d/t$.

Exfoliated clay modified poly(ethylene terephthalate) (PET) is one of the more prevalent nanocomposites investigated by both academic and industrial laboratories for barrier applications [197], [198] and [199]. In situ polymerized PET-exfoliated clay composites were noted to show a 2-fold reduction in permeability with only 1 wt% clay versus the control PET [197]. PET-exfoliated clay composites also prepared via in situ polymerization using a clay-supported catalyst exhibited a 10 to 15-fold reduction in O2 permeability with 1–5 wt% clay [198]. The moisture vapor transmission, however, did not show any significant change. Exfoliated clay modified chlorobutyl rubber showed decreased diffusion for several organic chemicals suggesting utility for chemical protective gloves/clothing [200]. Exfoliated clay added to polyamide 6/polyolefin (polyethylene or polypropylene) blends yielded an improved barrier to styrene permeation for melt blown films [201]. It was noted that the polymer blend nanocomposite was a better barrier than the control polyamide nanocomposite.

While most of the papers investigating barrier properties incorporate low levels of exfoliated clay, a novel approach employed producing a self-supporting clay fabric film followed by infiltration with an epoxy resin/amine hardener mixture and polymerization [202]. The resultant semitransparent nanocomposite film contained up to 77% volume fraction clay with an oxygen permeability 2–3 orders of magnitude lower than the control epoxy.

The concept of mixed matrix membranes involving molecular sieve inclusions in a polymer film to enhance the permselectivity properties for membrane separation was developed by Koros et al. [203] to address the limits imposed by upper bound limits typically observed with polymer membranes [204]. These inclusions (carbon molecular sieves, zeolite structures) need to be at nanolevel dimensions as the dense layer thickness of commercial membranes is in the range of 100 nm. This approach has shown promise in exceeding the noted upper bound in various studies [205] and [206]. The addition of silica nanoparticles to poorly packing polymer membranes (specifically poly(4-methyl-2-pentyne)) has been shown to yield even poorer packing and higher free volume [44]. This leads to increased permeability for larger organic molecules and a selectivity reversal for mixtures of these molecules with smaller molecules (e.g., n-butane/methane). This indicates that the separation process has changed from molecular sieving expected at low free volume and small void diameters to surface diffusion as the free volume and void diameters increase. The addition of nanoparticle TiO2 to poly(trimethyl silylpropyne) (also a high free volume polymer with poor chain packing) showed a decrease in gas permeability up to 7 vol% TiO2 with increasing permeability and higher free volume observed above 7 vol% loading [207].

Flammability Resistance

Increased flammability resistance has been noted as an important property enhancement involving nanoplatelet/nanofiber modification of polymeric matrices. While the specific reasons for this are under continuing investigation, a qualitative explanation observed in many studies involves the formation of a stable carbon/nanoplatelet or nanofiber surface. This surface exhibits analogous characteristics to intumescent coatings whereby the resultant "char" provides protection to the interior of the specimen by preventing continual surface regeneration of available fuel to continue the combustion process. The primary advantage noted with nanofiller incorporation is the reduction in the maximum heat release rate (determined by cone calorimetry) [137] and [208]. While significant reductions can be observed in the maximum heat release rates, the total heat release remains constant with nanofiller addition. The relevance of reducing the maximum heat release rate is to minimize the flame propagation to adjacent areas in the range of the ignited material (dimension range of meter(s)). The flammability improvements for nanofiller addition are less advantageous when the more common empirical regulatory (pass/fail) flammability tests are conducted (UL94, ASTM flammability tests) [137], [208] and [209]. In specific cases, the nanoparticle addition can result in reduced flammability rating due to the melt viscosity increase preventing dripping as a mechanism of flame extinguishment (e.g., change UL94 rating from V-2 to HB) [209]. The primary advantage for nanofiller addition for these tests generally involves reduction in the flame retardant additives that need to be incorporated to pass the specific test [137], [209] and [210]. This has been observed in various nanoparticle modified composites including exfoliated clay with halogen-based fame retardants/Sb2O3[211] and EVA (ethylene-vinyl acetate copolymer) nanocomposites with magnesium hydroxide nanoparticles and microcapsulated red phosphorus [212].

The majority of the flame retardant studies on nanofiller incorporation in polymers involves exfoliated clay. Studies involving polyamide 6 [213] and [214] and polypropylene [215] yielded similar observations with reduced peak heat release rate but no change in the total heat release with exfoliated clay addition. This is illustrated in Fig. 14 for a generalized data for nanofiller modified polymers. While the curve position and shape will vary for different polymer matrix materials and nanofiller incorporation, the generalized behavior of decreased peak heat release rate with basically no change in the overall heat release (area under the curve) is very typical. The surface characteristics during and after forced combustion show that incomplete surface coverage will lead to poorer flammability resistance and can be related to low nanofiller level, low aspect ratio, poor dispersion and/or agglomeration during combustion. This generalized behavior, illustrated in Fig. 15, is true for exfoliated clay as well as carbon nanotubes as discussed below.

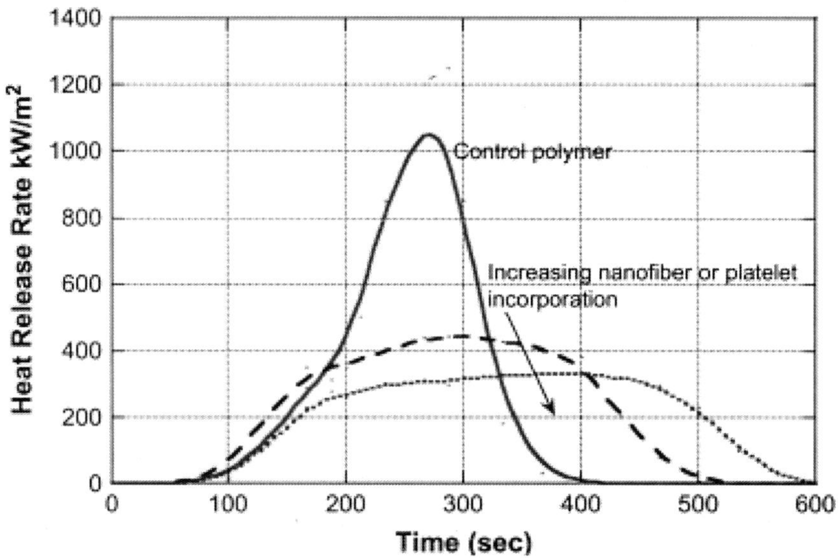

Figure 14. Generalized behavior for nanofiber or platelet modified polymers in the cone calorimetry heat release rate test.

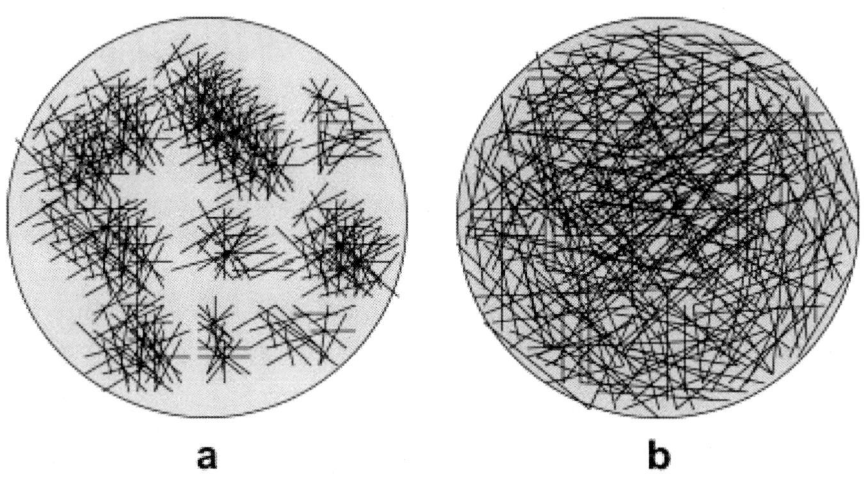

Figure 15. Surface structure of nanocomposites during/after forced combustion: (a) incomplete surface coverage due to low nanofiber level, agglomeration during combustion and/or low aspect ratio; (b) desired surface dispersion during/after forced combustion.

Studies involving carbon nanotubes have also shown the decrease in the peak heat release rate with no change in the total heat release [216] and [217] with effectiveness equal to or better than exfoliated clay. The level of dispersion of the carbon nanotubes in the polymer matrix was shown to be an important variable [216]. Upon combustion, the surface layer was enriched with a protective nanotube network providing a thermal and structural barrier to the combustion process. Continuity of the network (as illustrated in Fig. 15) was important to achieve optimum performance as very low levels of nanotube incorporation or poor dispersion did not allow a continuous surface network during the combustion process. It is noted that the incorporation of nanoclay and carbon nanotubes often results in slightly earlier ignition than the unmodified polymer presumably due to the increased thermal conductivity. However, at the later stages of combustion, the reinforcement of the char layer provides a stable thermal barrier preventing regeneration of polymer at the surface available for rapid combustion.

Polymer Blend Compatibilization

A primary mechanism in compatibilization of phase separated polymer blends involves lowering the interfacial tension between the phases and preventing coalescence of the particles during melt processing. This can be achieved by addition of graft or block copolymers with constituents equal to or compatible with the blend components. It has been observed in many cases that the addition of nanoparticles (particularly exfoliated clay) can also prevent the coalescence retaining improved dispersion after shear mixing. Specific examples involving exfoliated clay compatibilization include polycarbonate/poly(methyl methacrylate) [218], poly(2,6-dimethyl-1,4-phenylene oxide)/polyamide 6 [219], polyamide 6/ethylene–propylene rubber [220], polystyrene/poly(methyl methacrylate) [221] and poly(vinylidene fluoride)/polyamide 6 blends [222]. An example of exfoliated clay compatibilization of a poly(methyl methacrylate)/polystyrene (70/30 by weight) blend is illustrated in Fig. 16[221]. After shearing and annealing above the blend component Tgs, the blend containing exfoliated clay shows the ability to resist coalescence. Nanoscale SiO2 particle compatibilization was noted for polystyrene/polypropylene blends where a significant reduction in the polystyrene phase dimensions was observed [223]. The compatibilization was hypothesized to be due to increased viscosity retarding coalescence.

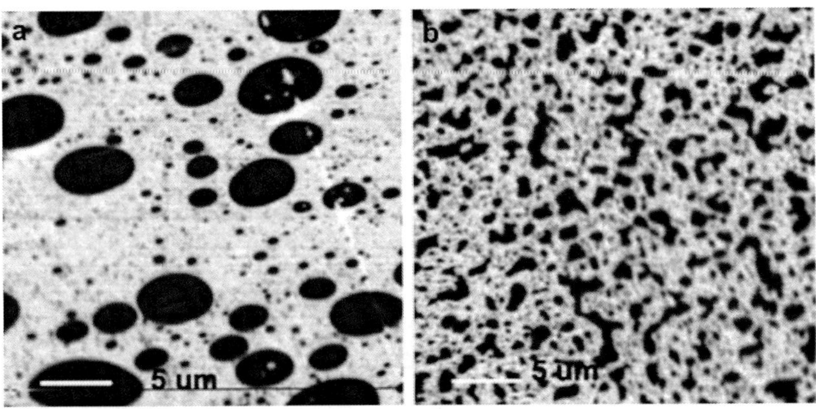

Figure 16. Scanning transmission X-ray microscopic images (30 μm × 30 μm) of PS/PMMA blends annealed at 190 °C for 14 h (taken at 285.2 eV, the adsorption energy of PS, PS is dark): (a) PS/PMMA (30/70), (b) PS/PMMA/Closite 6A (27/63/10). Reproduced with permission of Ref. [221], copyright (2006) American Chemical Society.

Several additional hypotheses have been noted in the literature. One explanation is that the nanoparticles concentrate at the interface preventing coalescence by a barrier-type mechanism. Another hypothesis notes that both polymers bound by physical [224] or chemical interactions [221] on the nanoparticle will concentrate at the interface similar to a block or graft copolymer comprising both components of the blend. In the case of physical interactions, it was noted that for a polypropylene/polystyrene blend both polymers were intercalated into the clay gallery with the polymer chain extending outside the particle [224]. The nanoparticle size is important for both explanations and offers an advantageous property not achievable at higher length scales.

Biomedical Applications

The applicability of polymer nanotechnology and nanocomposites to emerging biomedical/biotechnological applications is a rapidly emerging area of development of which this discussion can only briefly cover. One area of intense research involves electrospinning for producing bioresorbable nanofiber scaffolds for tissue engineering applications. This might be construed as a nanocomposite as the resultant scaffold allows for cell growth yielding a unique composite system. Another area also involving nanofibers is the utilization of electrically conducting nanofibers based on conjugated polymers for regeneration of nerve growth in a biological living system.

Nanoparticle silver, silver oxide and silver salts have been incorporated into polymer matrices to provide antimicrobial/ biocidal activity [225], [226] and [227]. Nanoscale silver was shown to be a much better antimicrobial additive to nylon 6 than microscale silver particles [227] due to the much higher rate of silver ion release. Nanosilver (5–50 nm) at levels of 0.1–1.0 wt% in poly(methyl methacrylate) bone cement exhibited high antibacterial activity for joint arthroplasty utility [228] without the cytotoxicity of silver salts.

Polymer nanocomposites based on hydroxyapatite ($Ca_{10}(PO_4)_6(OH)_2$) have been investigated for bone repair and implantation [229]. Hydroxyapatite, a major constituent of hard tissue, exhibits undesirable mechanical properties if directly employed thus polymer-based matrix composites are desired. Biodegradation of the matrix is also desired to allow infiltration of new bone growth at the repair site. Often natural polymers (polysaccharides, polypeptides, collagen, chitosan) or synthetic biodegradable polymers are employed as the matrix in these studies. Collagen derived gelatin [230] and poly-2-hydroxyethylmethacrylate/poly(ε-caprolactone) [231] nanocomposites based on hydroxyapatite are examples of systems studied for bone repair systems.

Electrospinning of biodegradable polymer solutions is a popular method to produce nanofiber scaffolds for tissue engineering applications. Poly(L-lactic acid)/exfoliated montmorillonite clay/ salt solutions were electrospun followed by salt leaching/gas foaming [232]. The resultant scaffold structure contained both nano- and micro-sized pores offering a combination of cell growth and blood vessel invasion micro-dimensions along with nanodimensions for nutrient and metabolic waste transport.

Polymer matrix nanocomposites have been proposed for drug delivery/release applications. The addition of nanoparticles can provide an impediment to drug release allowing slower and more controlled release, and reduced swelling [233] and improved mechanical integrity [234] of hydrogel-based nanocomposites. Iron oxide nanoparticles have been investigated for various applications including drug delivery, magnetic resonance imaging contrast enhancement, immunoassay and cellular therapy. These investigations often employ magnetite (Fe_3O_4) dispersed in a polymeric microsphere or microcapsule involving biodegradable and/or natural polymers [235] and [236]. Poly(L-lysine) microspheres containing magnetic nanoparticles prepared by coacervation were prepared and characterized for potential use in targeted drug delivery applications [237]. Iron and cobalt nanoparticles encapsulated in polydimethylsiloxane have been noted for treating retinal detachment disorders [238] and [239]. A novel imprint lithography method based on a crosslinked perfluoroether templating mold has been employed to fabricate uniform micro- and nanoparticles from organic polymers including biodegradable polymers such as poly(lactic acid) [240]. Inclusion of inorganic nanoparticles such as discussed above has potential for a diverse range of biomedical applications using this method.

Fuel Cell Applications

Fuel cell applications involve polymers in the proton exchange membrane, binder for the electrodes and matrix for bipolar plates. The electrodes typically comprise carbon black particles (0.5–1.0 μm) with Pt catalyst particles of 2–5 nm and a polymeric binder (usually Nafion®). Platinum nanoparticles deposited onto single-walled carbon nanotubes with Nafion® as a binder were shown to give improved performance over the conventional carbon black-based electrodes [241]. Nanoparticle incorporation in the proton exchange membrane has been noted in numerous publications to improve mechanical properties as well as to enhance proton conductivity. Additionally with direct methanol fuel cells, nanoparticles have been incorporated to reduce methanol crossover [242].

Heteropolyacids (HPA) (e.g., $H_3PW_{12}O_{40}$; $H_3PMo_{12}O_{40}$) have been added to proton exchange membranes to yield improved proton conductivity at higher temperatures while retaining good mechanical properties[243], [244], [245] and [246]. The particle size of the HPA inclusions was generally in the nanorange. Silica nanoparticle inclusion in proton exchange membranes gave lower methanol crossover in several studies[247] and [248]. Zirconium phosphate [249], zirconium hydrogen phosphate [250] and TiO_2[251] nanoparticle incorporation in proton exchange membranes exhibited promise in direct methanol fuel cells. Nanoclay modified Nafion® (laponite [252] and montmorillonite [253]) membranes have been reported to offer improvements over the unmodified controls. Sulfonated carbon nanofibers incorporated into a sulfonated EPDM proton exchange membrane yielded an order of magnitude improvement in the proton conductivity achieving performance comparable to the state-of-the-art Nafion® 117 [254].

Electrical/electronics, Optoelectronics, and Sensors

Nanotechnology is deeply embedded in the design of advanced devices for electronic and optoelectronic applications. The dimensional scale for electronic devices has now entered the nano-range. The utility of polymer-based nanocomposites in these areas is quite diverse involving many potential applications as well as types of nanocomposites. One specific nanocomposite type receiving considerable interest involves conjugated polymers and carbon nanotubes. A recent review of this area notes a litany of potential applications including photovoltaic (PV) cells and photodiodes, supercapacitors, sensors, printable conductors, light emitting diodes (LEDs) and field effect transistors [255]. This paper can only briefly discuss the nanocomposite technology applied to this broad field.

The electrical conductivity of carbon nanotubes in insulating polymers has also been a topic of considerable interest. The potential applications include electromagnetic interference shielding, transparent conductive coatings, electrostatic dissipation, supercapacitors, electromechanical actuators and various electrode applications [256] and [257]. The percolation threshold for electrical conductivity of epoxy composites containing multi-wall carbon nanotubes was found to be 0.0025 wt% MWCNT [258] considerably lower than nanoscale dispersed carbon black particles. The threshold conductivity of single-walled carbon nanotubes in epoxy composites was noted to be a function of the SWCNT type with values as low as 0.00005 vol fraction[259]. Water dispersed carbon black particles

(42 nm) added to acrylic emulsions yielded electrical conductivity percolation levels as low as 1.5 vol% in dried films [260]. In this system, the carbon black particles concentrate at the interface between the emulsion particles during drying yielding a percolation network. The modulus of the emulsion system chosen as the matrix was noted to be an important variable with higher modulus (higher T_g) yielding lower threshold percolation values.

The basic device structure for organic/polymeric PV and LED devices is shown in Fig. 17. The major difference between the two devices is the light emitting layer and the light harvesting layer. While all the potential layers are noted, actual devices often do not contain all the layers. Polymer-based nanocomposites are most relevant for the anode, hole injection layer, light emitting layer and light harvesting layer. In the case where flexible transparent substrates are desired, nanocomposite or nanolayered combinations of polymer and barrier nanomaterials can be employed to provide barrier properties to oxygen and water permeation.

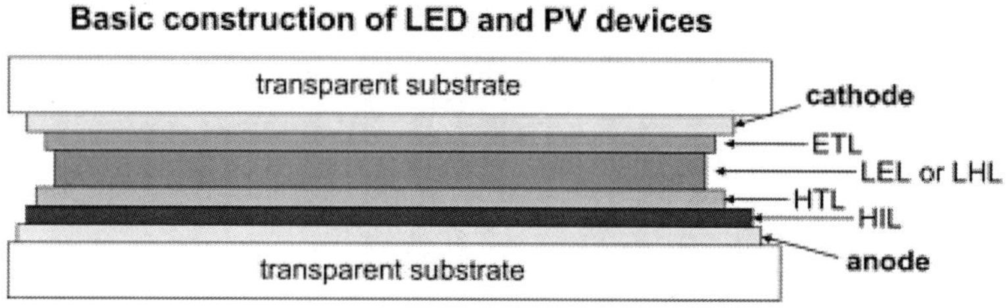

Figure 17. Generalized diagram of LED and PV device construction.

Silicon-based photovoltaic (PV) devices offer high efficiency, excellent stability and proven commercial utility. Organic/polymer-based PV devices offer the potential for lower cost and more flexible manufacture but need significant improvements in both efficiency and long term stability. Conjugated polymers are often employed in the organic-based device light harvesting layer but have limitations in charge transport. Combinations of conjugated polymer with inorganic semiconductors have been proposed as a resolution to this deficiency. Cadmium-selenium, CdSe, nanorod incorporation in poly(3-hexylthiophene) yielded power conversion efficiency of 1.7% [261]. Silicon solar cells, however, typically deliver power conversion efficiencies of 10% or greater. Layer-by-layer assembly of functionalized poly(phenylene vinylene) and CdSe nanoparticle composites yielded uniform thin films with a power conversion efficiency of 0.71% [262]. A functionalized polystyrene containing an electroactive carbazole pendant group and an amine salt pendant group capable of electrostatic interaction with CdTe was described for potential PV applications [263]. Poly(3-hexylthiophene)–ZnO nanofiber composites exhibited a power conversion efficiency of 0.53% noted to be significantly better than the analogous bilayer structure of the noted components [264]. Spherical ZnO nanoparticle (20–40 nm) incorporation into poly(3-hexylthiophene) was ineffective at low levels of incorporation (20–30 wt%) but large improvements were observed at 50–60 wt% addition yielding a power conversion efficiency increase of 35 times to a value of 0.42% [265]. SWCNT incorporation into poly(3-octylthiophene) increased the short circuit current by two orders of magnitude and improved the important fill factor [266]. Ink jet printing is a potential method for producing low cost, high volume PV and LED devices. The ink jet printing of the nanocomposite of poly(vinyl alcohol) and CdTe for these devices was demonstrated by Tekin et al. [267].

The primary material presently employed for the hole injection layer (HIL) is PEDOT:PSS (poly(3,4-ethylene dioxythiophene):poly(styrene sulfonic acid)) but has limitations in life-time performance. The addition of silver nanoparticles (3–6 nm) to PEDOT:PSS resulted in maximum device luminance at 20 wt% [268]. The addition of nickel nanoparticles (30 nm) into the PEDOT:PSS HIL was also shown to improve device performance attributed to the improved hole current [269]. Gold nanoparticle (5–10 nm) incorporation into poly(9,9′-dioctylfluorene) (LEL layer) gave higher external quantum efficiencies and improved oxygen stability compared to the unmodified polyfluorene [270] and [271]. Poly(9,9′-dioctylfluorene)-layered silicate nanocomposites were also demonstrated to improve device stability as well as offering higher quantum efficiencies over the unmodified polyfluorene [272]. Poly(phenylene) with polyhedral oligomeric silsesquioxane (POSS) side groups inhibited interchain interaction leading to improved fluorescence quantum yields in solution and photoluminescent thermal stability in bulk [273].

Exfoliated graphite sheets (graphene) have been only recently investigated in nanocomposites [274] at least partly due to the lack of methods to achieve high levels of exfoliation [275] and [276]. It was noted that the electrical conductivity for polystyrene–graphene nanocomposites had a percolation threshold of 0.1 vol% rivaling carbon nanotubes [274]. The electrical conductivity at 1 vol% was 0.1 S/cm. The emerging interest of graphene in electronics applications parallels that which occurred with carbon nanotube discovery. Graphene, in sheet form, may offer promise in replacing silicon as Moore's law limits are achieved [277].

Conjugated polymers with various nanoscale filler inclusions have been investigated for sensor applications including gas sensors, biosensors and chemical sensors. The nanofillers employed include metal oxide nanowires, carbon nanotubes, nanoscale gold, silver, nickel, copper, platinum and palladium particles [278]. With carbon nanotubes, the electrical resistance was found to be significantly changed by exposure to specific gases such as NO2 and NH3[279]. A nanocomposite of SWCNT/polypyrrole yielded a gas sensor sensitivity similar to SWCNT alone [280]. The sensing capability of these nanocomposites can be based on conductivity changes due to gas or chemical interactions with either the nanofiller or the conjugated polymer, pH changes, electrochromic or electro-optical property changes, catalytic activity, chemiluminescent property or biological recognition. Examples of sensors for dopamine detection include a poly(anilineboronic acid)/carbon nanotube composite [281] and a polyaniline/gold composite hollow sphere system [282]. A review of conjugated polymer nanocomposites employed as sensors has been recently published [278].

COMMERCIAL APPLICATIONS OF POLYMER-BASED NANOCOMPOSITES

A question often posed is "with all the interest and associated large R&D expenditures in nanotechnology (including polymer nanocomposites) why is there not more commercial impact?" Often major discoveries take several decades to reach large commercial impact (polyethylene, carbon fiber composites as examples) as the cost/performance variables upon discovery are outside the realm of commodity utility and additional advances are required to achieve an economically competitive position. The commodities of today were the specialties of the past and even the automobile was a specialized article of commerce for several decades. Such is the case with the technologies being developed with polymer-based nanocomposites. While many of the applications being commercialized today will remain specialties, there are areas where the specialty polymer nanocomposites of today will be the commodities of the future. Examples of commercial polymer based nanocomposites are listed in Table 4.

Table 4. Examples of nanocomposite commercial utility

Polymer matrix	Nanoparticle	Property improvement	Application	Company and/or product trade name
Polyamide 6	Exfoliated clay	Stiffness	Timing belt cover: automotive	Toyota/Ube
TPO (thermoplastic polyolefin)	Exfoliated clay	Stiffness/strength	Exterior step assist	General Motors
Epoxy	Carbon nanotubes	Strength/stiffness	Tennis rackets	Babolat
Epoxy	Carbon nanotubes	Strength/stiffness	Hockey sticks	Montreal: Nitro Hybtonite®
Polyisobutylene	Exfoliated clay	Permeability barrier	Tennis balls, tires, soccer balls	InMat LLC
SBR, natural rubber, polybutadiene	Carbon black (20–100 nm: primary particles)	Strength, wear and abrasion	Tires	Various
Various	MWCNT	Electrical conductivity	Electrostatic dissipation	Hyperion
Unknown	Silver	Antimicrobial	Wound care/ bandage	Curad®

Nylon MXD6, PP	Exfoliated clay	Barrier	Beverage containers, film	Imperm™: Nanocor
SBR rubber	Not disclosed	Improved tire performance in winter	Winter tires	Pirelli
Natural rubber	Silver	Antimicrobial	Latex gloves	
Various	Silica	Viscosity control, thixotropic agent	Various	
Polyamides nylon 6, 66, 12	Exfoliated clay	Barrier	Auto fuel systems	Ube

The most publicized application for polymer nanocomposites was an automotive application by Toyota for a timing belt cover. The utilization of exfoliated clay reinforcement of TPO (thermoplastic polyolefin) came later utilized by General Motors for an exterior step assist for another automotive application. Obviously, there are many more applications involving exfoliated clay reinforcement, however, many of these have not been publicized such as those noted above.

One of the initial uses for exfoliated clay in barrier applications involved a 20 μm coating on the interior of a tennis ball to prevent depressurization. The product was developed by InMat LLC and introduced in 2001. Sports equipment was one of the initial areas where carbon fiber composites were commercialized. This is also true for carbon nanotubes where the carbon nanotubes (at low levels) reinforce the epoxy matrix of the carbon fiber composite in specialty tennis rackets and hockey sticks. In such applications, performance overrides the economic disadvantages of the expensive carbon nanotube inclusion.

Comments on the Future of Polymer Matrix Based Nanocomposites

An area where nanocomposites could achieve a dramatic commercial prominence is in advanced composites. Carbon fiber reinforced composites have a limit on the achievable properties (particularly in cross-ply composites) due to the low modulus and strength of the matrix phase. Modification of the matrix phase with carbon nanotubes at the lower scale of dimensions and carbon nanofibers at a higher dimensional scale would allow for significant increases in the modulus and strength contributions of the matrix to the overall composite properties. While this would offer some improvement in unidirectional composites, it could be dramatic in the case of cross-ply composites which are the major type of composite structure utilized in advanced composite applications. This concept is under present consideration and could allow a step change in the advanced composite field. The basic concept has already been commercially employed in specialty sports equipment (tennis rackets and hockey sticks). These improvements are key to future aircraft and wind energy turbine applications. This approach is analogous to naturally occurring composite structures where the hierarchical construction method employs several dimensional scales beginning at the nanolevel. The same concept is also relevant to the more commodity reinforced composites where exfoliated clay could be added to the matrix for unsaturated polyester–fiberglass composites or other fiberglass reinforced matrix polymers.

As the secrets of nature's methodology to optimize material properties by nanolevel construction are unlocked (biomimetics), translation of these findings to polymer nanocomposites should allow for further advances. Nanostructured surfaces have been noted to yield superhydrophobic character (lotus leaf) and exceptional adhesion (gecko foot). The confluence of the biological and polymer material science disciplines often involves the design of nanoscale polymeric blends and composite systems to mimic the biological systems.

Carbon nanotube or exfoliated graphite (graphene) offers substantial opportunities in the electrical/electronics/optoelectronics areas as well as potential in specific emerging technologies. One specific area would be replacement of ITO as a transparent conductor for lower cost and flexible devices. Carbon nanotube sheets have been proposed [283] and the potential for carbon nanotube-conjugated polymer composites would be of interest if sufficient electrical conductivity can be obtained (>10^3 S/cm). The potential of low cost graphene production is yet to be realized and large scale utility will be awaiting the synthetic breakthrough.

An additional area not discussed in detail in this review

involves the importance of morphology control which includes both dispersion and alignment. These issues have been discussed in many of the papers already cited in this review as well as in recent reviews [156] and [284]. Obtaining the optimum properties for nanocomposites will usually require excellent dispersion of the nanoparticles. The tendency for nanoparticles (including platelets and fibers of nanoscale dimensions) to coalesce into macrosize agglomerates can seriously impact the achievable properties. In specific cases, excellent isotropic dispersion may not be the desired morphology but rather a hierarchical morphology that offers unique properties such as those observed with percolation pathways to maximize electrical conductivity or patterned morphology to achieve novel optical or electronic properties [284].

REFERENCES

1. Mark JE, Jiang CY, Tang MY. Macromolecules 1984;17:2613–6.
2. Wilkes GL, Orler B, Huang H. Polym Prep 1985;26:300–1.
3. Wen J, Wilkes GL. Chem Mater 1996;8:1667–81.
4. Kojima Y, Usuki A, Kawasumi M, Okada A, Kurauchi T, Kamigaito O. J Polym Sci Part A Polym Chem 1993;31:983–6.
5. Okada A, Fukushim Y, Kawasumi M, Inagaki S, Usuki A, Sugiyama S, et al. US Patent 4,739,007; 1988.
6. Kawasumi M. J Polym Sci Part A Polym Chem 2004;42:819–24.
7. Loo YL, Register RA, Ryan AJ. Phys Rev Lett 2000;84:4120–3.
8. Loo YL, Register RA, Ryan AJ. Macromolecules 2002;35:2365–74.
9. Santana OO, Mu¨ ller AJ. Polym Bull 1994;32:471–7.
10. Woo E, Huh J, Jeong YG, Shin K. Phys Rev Lett 2007;98:136103.
11. Shin K, Woo E, Jeong YG, Kim C, Huh J, Kim KW. Macromolecules 2007;40: 6617–23.
12. Wu H, Wang W, Yang H, Su Z. Macromolecules 2007;40:4244–9.
13. Di Maio E, Iannace S, Sorrentio L, Nicolais L. Polymer 2004;45:8893–900.
14. Zhang QX, Yu ZZ, Yang M, Ma J, Mai YW. J Polym Sci Part B Polym Phys 2003; 41:2861–9.
15. Mu B, Wang Q, Wang H, Jian L. J Macromol Sci Part B Phys 2007;46:1093–104.
16. Nam JY, Ray SS, Okamoto M. Macromolecules 2003;36:7126–31.
17. Lincoln DM, Vaia RA, Krishnamoorti R. Macromolecules 2004;37:4554–61.
18. Li L, Li CY, Ni C, Rong L, Hsiao B. Polymer 2007;48:3452–60.
19. Phang IY, Pramoda KP, Liu T, He C. Polym Int 2004;53:1282–9.

20. Wu D, Zhou C, Fan X, Mao D, Bian Z. Polym Polym Compos 2005;13:61–71.
21. Bilotti E, Fischer HR, Peijs T. J Appl Polym Sci 2008;107:1116–23.
22. Xu D, Wang Z. Polymer 2008;49:330–8.
23. Homminga D, Goderis B, Dolbnya I, Reynaers H, Groeninckx G. Polymer 2005;46:11359–65.
24. Chen EC, Wu TM. J Polym Sci Part B Polym Phys 2008;46:158–69.
25. Kim B, Lee SH, Lee D, Ha B, Park J, Char K. Ind Eng Chem Res 2004;43: 6082–9.
26. Harrats C, Groeninckx G. Macromol Rapid Commun 2008;29:14–26.
27. Gue´ rin G, Prud'homme RE. J Polym Sci Part B Polym Phys 2007;45:10–7.
28. Rittigstein P, Torkelson JM. J Polym Sci Part B Polym Phys 2006;44: 2935–43.
29. Pluta M, Jeszka JK, Boiteux G. Eur Polym J 2007;43:2819–35.
30. Lee KJ, Lee DK, Kim YW, Choe WS, Kim JH. J Polym Sci Part B Polym Phys 2007;45:2232–8.
31. Xu H, Yang B, Wang J, Guang S, Li C. J Polym Sci Part A Polym Chem 2007;45: 5308–17.
32. Ramasundaram SP, Kim KJ. Macromol Symp 2007;249–250:295–302.
33. Huang JC, He CB, Xiao Y, Mya KY, Dai J, Siow YP. Polymer 2003;44:4491–9.
34. Pham JQ, Mitchell CA, Bahr JL, Tour JM, Krishanamorrti R, Green PF. J Polym Sci Part B Polym Phys 2003;41:3339–45.
35. Bo¨hning M, Goering H, Hao N, Mach R, Oleszak F, Scho¨ nhals A. Rev Adv Mater Sci 2003;5:155–9.
36. Xu W, Ge M, Pan WP. J Therm Anal Calorim 2004;78:91–9.
37. Fragiadakis D, Pissis P. J Non-Cryst Solids 2007;353:4344–52.
38. Shi X, Gan Z. Eur Polym J 2007;43:4852–8.
39. Huskic´ M, Zˇ igon M. Eur Polym J 2007;43:4891–7.
40. Yuen SM, Ma CCM, Lin YY, Kuan HC. Compos Sci Technol 2007;67:2564–73.
41. Uthirakumar P, Nahm KS, Hahn YB, Lee YS. Eur Polym J 2004;40:2437–44.
42. Sun Y, Luo Y, Jia D. J Appl Polym Sci 2008;107:2786–92.
43. Chang JH, Mun MK, Kim JC. J Appl Polym Sci 2007;106:1248–55.
44. Merkel TC, Freeman BD, Spontak RJ, He Z, Pinnau I, Meakin P, et al. Science 2002;296:519–22.
45. Kim JH, Koros WJ, Paul DR. Polymer 2006;47:3094–103 and 3104-3111.
46. Huang Y, Paul DR. J Polym Sci Part B Polym Phys 2007;45:1390–8.
47. Van Olphen H. An introduction to clay colloid chemistry. New York: Interscience; 1963.
48. LeBaron PC, Wang Z, Pinnavaia TJ. Appl Clay Sci 1999;15:11–29.
49. Pinnavaia TJ, Beall GW, editors. Polymer–clay nanocomposites. New York: John Wiley & Sons; 2000.
50. Yariv S, Cross H, editors. Organo-clay complexes and interactions. New

York: Marcel Dekker; 2002.

51. Ray SS, Okamoto M. Prog Polym Sci 2003;28:1539–641.
52. Mai Y, Yu Z, editors. Polymer nanocomposites. Cambridge: Woodhead; 2006.
53. Hussain F, Hojjati M. J Compos Mater 2006;40(17):1511–65.
54. Hunter DL, Kamena KW, Paul DR. MRS Bull 2007;32:2806.
55. Xie W, Gao Z, Liu K, PanW-P, Vaia R, Hunter D, et al. Thermochim Acta 2001; 367–368:339–50.
56. Xie W, Gao Z, Pan W-P, Hunter D, Singh A, Vaia R. Chem Mater 2001;13: 2979–90.
57. Xie W, Xie R, Pan W-P, Hunter D, Koene B, Tan L-S, et al. Chem Mater 2002; 14:4837–45.
58. Fornes TD, Paul DR. Polymer 2003;44:4993.
59. Paul DR, Zeng QH, Yu AB, Lu GQ. J Colloid Interface Sci 2005;292:462.
60. Heinz H, Vaia RA, Krishnamoorti R, Farmer BL. Chem Mater 2007;19:59–68.
61. Ploehn HJ, Liu C. Ind Eng Chem Res 2006;45:7025–34.
62. Fukushima Y, Inagaki S. J Inclusion Phenom 1987;5:473–82.
63. Kojima Y, Usuki A, Kawasumi M, Okada A, Kurauchi T, Kamigaito O. J Polym Sci Part A Polym Chem 1993;31:1755–8.
64. Biasci L, Aglietto M, Ruggeri G, Ciardelli F. Polymer 1994;35(15):3296–309.
65. Huang X, Lewis S, Brittain WJ. Macromolecules 2000;33:2000–4.
66. Zhou Q, Fan X, Xia C, Mays J, Advincula R. Chem Mater 2001;13:2465–7.
67. Albrecht M, Ehrler S, Mu¨ hlebach A. Macromol Rapid Commun 2003;24: 382–7.
68. Kiersnowski A, Piglowski J. Eur Polym J 2004;40:1199–207.
69. Goldberg HA, Feeney CA, Karim DP, Farrell M. Rubber World 2002;226:1–17. see also p. 20 and 37.
70. Takahashi S, Goldberg HA, Feeney CA, Karim DP, Farrell M, O'Leary K, et al. Polymer 2006;47:3083–93.
71. Vaia RA, Ishii H, Giannelis EP. Chem Mater 1993;5:1694.
72. Vaia RA, Teukolsky RK, Giannelis EP. Chem Mater 1994;6:1017–22.
73. Vaia RA, Jandt KD, Kramer EJ, Giannelis EP. Macromolecules 1995;28: 8080–5.
74. Giannelis EJ. Adv Mater 1996;8(1):29.
75. Vaia RA, Jandt KD, Kramer EJ, Giannelis EP. Chem Mater 1996;8:2628–35.
76. Vaia RA, Giannelis EP. Macromolecules 1997;30:7990–9.
77. Vaia RA, Giannelis EP. Macromolecules 1997;30:8000–9.
78. Lee JY, Baljon ARC, Lorin RF. J Chem Phys 1990;111(21):9754–60.
79. Manias E, Chen H, Krishnamoorti R, Genzer J, Kramer EJ, Giannelis EP. Macromolecules 2000;33:7955–66.
80. Anastasiadis SH, Karatasos K, Vlachos G. Phys Rev Lett 2000;84(5): 915–8.

81. Vaia RA, Giannelis EP. Polymer 2001;42:1281–5.
82. Cho JW, Paul DR. Polymer 2001;42:1083.
83. Dennis HR, Hunter DL, Chang D, Kim S, White JL, Cho JW, et al. Polymer 2001; 42:9513.
84. Fornes TD, Yoon PJ, Keskkula H, Paul DR. Polymer 2001;42:9929.
85. Fornes TD, Yoon PJ, Hunter DL, Keskkula H, Paul DR. Polymer 2002;43:5915.
86. Yoon PJ, Fornes TD, Paul DR. Polymer 2002;43:6727.
87. Fornes TD, Paul DR. Polymer 2003;44:3945.
88. Yoon PJ, Fornes TD, Paul DR. Polymer 2003;44:5323.
89. Yoon PJ, Hunter DL, Paul DR. Polymer 2003;44:5341.
90. Fornes TD, Yoon PJ, Paul DR. Polymer 2003;44:7545.
91. Fornes TD, Hunter DL, Paul DR. Polymer 2004;45:2321.
92. Shah RK, Paul DR. Polymer 2004;45:2991.
93. Chavarria F, Paul DR. Polymer 2004;45:8501.
94. Fornes TD, Paul DR. Macromolecules 2004;37:7698.
95. Bourbigot S, Vanderhart D, Gilman J, Stretz HA, Paul DR. Polymer 2004;45: 7627.
96. Hotta S, Paul DR. Polymer 2004;45:7639.
97. Ibanes C, David L, Seguela R, Rochas C, Robert G. J Polym Sci Part B Polym Phys 2004;42:2633–48.
98. Vlasveld DPN, Groenewold J, Bersee HEN, Mendes E, Picken SJ. Polymer 2005;46:6102–13.
99. Stretz HA, Paul DR, Li R, Keskkula H, Cassidy PE. Polymer 2005;46:2621.
100. Shah RK, Hunter DL, Paul DR. Polymer 2005;46:2646.
101. Stretz HA, Paul DR, Cassidy PE. Polymer 2005;46:3818.
102. Zeng QH, Yu AB, Lu GQ, Paul DR. J Nanosci Nanotechnol 2005;46:3818.
103. Lee H-S, Fasulo PD, Rodgers WR, Paul DR. Polymer 2005;46:11673.
104. Ahn YC, Paul DR. Polymer 2006;47:2830.
105. Shah RK, Paul DR. Macromolecules 2006;39:3327.
106. Lee H-S, Fasulo PD, Rodgers WR, Paul DR. Polymer 2006;47:3528.
107. Shah RK, Paul DR. Polymer 2006;47:4074.
108. Shah RK, Krishnaswarmy RK, Takahashi S, Paul DR. Polymer 2006;47: 6187.
109. Chavarria F, Paul DR. Polymer 2006;47:7760.
110. Stretz HA, Paul DR. Polymer 2006;47:8123.
111. Stretz HA, Paul DR. Polymer 2006;47:8527.
112. Shah RK, Cui L, Williams KL, Bauman B, Paul DR. J Appl Polym Sci 2006;102: 2980.
113. Shah RK, Kim DH, Paul DR. Polymer 2007;48:1047.
114. Cui L, Paul DR. Polymer 2007;48:1632.

115. Chavarria K, Nairn K, White P, Hill AJ, Hunter DL, Paul DR. J Appl Polym Sci 2007;105:2910.
116. Yoo Y, Shah RK, Paul DR. Polymer 2007;48:4867.
117. Kim DH, Fasulo PD, Rodgers WR, Paul DR. Polymer 2007;48:5308.
118. Kim DH, Fasulo PD, Rodgers WR, Paul DR. Polymer 2007;48:5960.
119. Cui L, Ma X, Paul DR. Polymer 2007;48:6325.
120. Chavarria K, Shah RK, Hunter DL, Paul DR. Polym Eng Sci 2007;47:1847.
121. Liu L, Qi Z, Zhu X. J Appl Polym Sci 1999;71:1133–8.
122. Heinemann J, Reichert P, Thomann R, Mu¨ lhaupt R. Macromol Rapid Commun 1999;20:423–30.
123. Garce´ s JM, Moll DJ, Bicerano J, Fibiger R, McLeod DG. Adv Mater 2000;12(23): 1835–9.
124. Varlot K, Reynaud E, Kloppfer MH, Vigier G, Varlet J. J Polym Sci Part B Polym Phys 2001;39:1360–70.
125. Reichert P, Hoffmann B, Bock T, Thomann R, Mu¨ lhaupt R, Friedrich C. Macromol Rapid Commun 2001;22(7):519–23.
126. Kim SW, Jo WH, Lee MS, Ko MB, Jho JY. Polymer 2002;34(3):103–11.
127. Ray SS, Yamada K, Ogami A, Okamoto M, Ueda K. Macromol Rapid Commun 2002;23:943–7.
128. Ray SS, Yamada K, Okamoto M, Fujimoto Y, Ogami A, Ueda K. Polymer 2003; 44:6633–46.
129. Ray SS, Okamoto K, Okamoto M. Macromolecules 2003;36:2355–67.
130. Yalcin B, Valladares D, Cakmak M. Polymer 2003;44:6913–25.
131. Yalcin B, Cakmak M. Polymer 2004;45:2691–710.
132. Konishi Y, Cakmak M. Polymer 2005;46:4811–26.
133. Hsieh AJ, Moy P, Beyer FL, Madison P, Napadensky E. Polym Eng Sci 2004; 44(5):825–37.
134. Weon J-I, Sue H-J. Polymer 2005;46:6325–34.
135. Vermogen A, Masenelli-Varlot K, Se´gue´ la R, Duchet-Rumeau J, Boucard S, Prele P. Macromolecules 2005;38:9661–9.
136. Kim Y, White JL. J Polym Sci 2005;96:1888–96.
137. Morgan AB. Polym Adv Technol 2006;17:206–17.
138. Masenelli-Varlot K, Vigier G, Vermogen A, Gauthier C, Cavaille´ JY. J Polym Sci Part B Polym Phys 2007;45:1243–51.
139. Picard E, Vermogen A, Ge´ rard J-F, Espuche E. J Membr Sci 2007;292:133–44.
140. Krishnamoorti R. MRS Bull 2007;32:341.
141. Fornes TD, Hunter DL, Paul DR. Macromolecules 2004;37:1793.
142. Tanaka G, Goettler LA. Polymer 2002;43:541–53.
143. Kawasumi M, Hasegawa N, Kato M, Usuki A, Okada A. Macromolecules 1997; 30:6333–8.

144. Hasegawa N, Kawasumi M, Kato M, Usuki A, Okada A. J Appl Polym Sci 1998; 67:87–92.

145. Kato M, Okamoto H, Hasegawa N, Tsukigase A, Usuki A. Polym Eng Sci 2003; 43(6):1312.

146. Hasegawa N, Usuki A. J Appl Polym Sci 2004;93:464–70.

147. Toth R, Coslanich A, Ferrone M, Fermeglia M, Pricl S, Miertus S, et al. Polymer 2004;45:8075–83.

148. Sza´ zdi L, Puka´nszky B, Fo¨ ldes E, Puka´nszky B. Polymer 2005;46:8001–10.

149. Davis CH, Mathias LJ, Gilman JW, Schiraldi DA, Shields JR, Trulove P, et al. J Polym Sci Part B Polym Phys 2002;40:2661–6.

150. Vaia RA, Liu W. J Polym Sci Part B Polym Phys 2002;40:1590–600.

151. Morgan AB, Gilman JW. J Appl Polym Sci 2003;87:1329–38.

152. Lincoln DM, Vaia RA, Wang Z-G, Hsiao BS, Krishnamoorti R. Polymer 2001; 42:9975–85.

153. Lincoln DM, Vaia RA, Wang Z-G, Hsiao BS. Polymer 2001;42:1621–31.

154. Vaia RA, Liu W, Koerner H. J Polym Sci Part B Polym Phys 2003;41:3214–36.

155. Justice RS, Schaefer DW, Vaia RA, Tomlin DW, Bunning TJ. Polymer 2005;46: 4465–73.

156. Schaefer DW, Justice RS. Macromolecules 2007;40(24):8501–17.

157. VanderHart DL, Asano A, Gilman JW. Chem Mater 2001;13:3796–809.

158. VanderHart DL, Asano A, Gilman JW. Chem Mater 2001;13:3781–95.

159. VanderHart DL, Asano A. Macromolecules 2001;34:3819–22.

160. Davis RD, Gilman JW, VanderHart DL. Polym Degrad Stab 2003;79: 111–21.

161. Ho DL, Briber RM, Glinka CJ. Chem Mater 2001;13:1923–31.

162. Schmidt G, Nakatani AI, Butler PD, Han CC. Macromolecules 2002;35: 4725–32.

163. Oshinski AJ, Keskkula H, Paul DR. Polymer 1996;37(22):4891–907.

164. Corte´ L, Leibler L. Polymer 2005;46:6360–8. 3202 D.R.

165. Lee KY, Paul DR. Polymer 2005;46:9064.

166. Lee KY, Kim KH, Jeoung SK, Ju SI, Shim JH, Kim NH, et al. Polymer 2007;48: 4174.

167. Mori T, Tanaka K. Acta Metall 1973;21:571.

168. Halpin JC, Kardos JL. Polym Eng Sci 1976;16(5):344.

169. Chow TS. J Polym Sci 1978;16:959–65.

170. Chow TS. J Polym Sci 1978;16:967–70.

171. Chow TS. J Mater Sci 1980;15:1873–88.

172. Hine PJ, Lusti HR, Gusev AA. Compos Sci Technol 2002;62:1445–53.

173. Lusti HR, Hine PJ, Gusev AA. Compos Sci Technol 2002;62:1927–34.

174. Van Es M, Xiqiao F, van Turnhout J, van der Giessen E. In: Al-Malaika S, Golovoy A, Wilkie CA, editors. Specialty polymer additives: principles and

applications. Oxford: Blackwell Science; 2001. p. 391–414.
175. Brune DA, Bicerano J. Polymer 2002;43:369–87.
176. Luo J-J, Daniel IM. Compos Sci Technol 2003;63:1607–16.
177. Zhu L, Narh KA. J Polym Sci Part B Polym Phys 2004;42:2391–406.
178. Tsai J, Sun CT. Compos Mater 2004;38(7):567–79.
179. Sharaf MA, Mark JE. Polymer 2004;45:3943–52.
180. Wang J, Pyrz R. Compos Sci Technol 2004;64:925–34.
181. Wang J, Pyrz R. Compos Sci Technol 2004;64:935–44.
182. Shepard PD, Golemba FJ, Maine FW. Adv Chem Ser 1973;134:41–51.
183. Sheng N, Boyce MC, Parks DM, Rutledge GC, Abes JI, Cohen RE. Polymer 2004; 45:487–506.
184. Hbaieb K, Wang QX, Chia YHJ, Cotterell B. Polymer 2007;48:901–9.
185. Sen S, Thomin JD, Kumar SK, Keblinski P. Macromolecules 2007;40:4059–67.
186. Laura DM, Keskkula H, Barlow JW, Paul DR. Polymer 2002;43:4673–87.
187. Krishnamoorti R, Giannelis EP. Macromolecules 1997;30:4097–102.
188. Medellin-Rodriguez FJ, Burger C, Hsiao BS, Chu B, Vaia R, Phillips S. Polymer 2001;42:9015–23.
189. Wagener R, Reisinger TJG. Polymer 2003;44:7513–8.
190. Chen B, Evans JRG. Scr Mater 2006;54:1581–5.
191. Manevitch OL, Rutledge GC. J Phys Chem B 2004;108:1428–35.
192. Nielsen LE. J Macromol Sci (Chem) 1967;A1:929–42.
193. Lape NK, Nuxoll EE, Cussler EL. J Membr Sci 2004;236:29–37.
194. Gusev AA, Lusti HR. Adv Mater 2001;13:1641–3.
195. Fredrickson GH, Bicerano J. J Chem Phys 1999;110:2181–8.
196. Bharadwaj K. Macromolecules 2001;34:9189–92.
197. Kim SH, Kim SC. J Appl Polym Sci 2007;103:1262–71.
198. Choi WJ, Kim HJ, Yoon KH, Kwon OH, Hwang CI. J Appl Polym Sci 2006;100: 4875–9.
199. Matayabas Jr JC, Turner SR. In: Pinnavaia TJ, Beall GW, editors. Polymer-clay nanocomposites. John Wiley & Sons Ltd.; 2000. p. 207–25; See also: Matayabas Jr JC, Turner SR, Sublett BJ, Connell GW, Gilmer JW, Barbee RB. US Patent 6,084,019, assigned to Eastman Chemical Corp.; July 4, 2000.
200. Sridhar V, Tripathy DK. J Appl Polym Sci 2006;101:3630–7.
201. Brule´ B, Flat JJ. Macromol Symp 2006;233:210–6.
202. Triantafyllidis KS, LeBaron PC, Park I, Pinnavaia TJ. Chem Mater 2006;18: 4393–8.
203. Zimmerman CM, Singh A, Koros WJ. J Membr Sci 1997;137:145–54.
204. Robeson LM. J Membr Sci 1991;62:165–85.
205. Husain S, Koros WJ. J Membr Sci 2007;288:195–207.

206. Chung TS, Jiang LY, Li Y, Kulprathipanja S. Prog Polym Sci 2007;32:483–507.
207. Matteucci S, Kusuma VA, Sanders D, Swinnea S, Freeman BD. J Membr Sci 2008;307:196–217.
208. Bourbigot S, Duquesne S, Jama C. Macromol Symp 2006;233:180–90.
209. Schartel B, Bartholmai M, Knoll U. Polym Adv Technol 2006;17:772–7.
210. Nazare S, Kandola BK, Horrocks AR. Polym Adv Technol 2006;17:294–303.
211. Zanetti M, Camino G, Canavese D, Morgan AB, Lamelas FJ, Wilkie CA. Chem Mater 2002;14:189–93.
212. Lv JP, Liu WH. J Appl Polym Sci 2007;105:333–40.
213. Dasari A, Yu ZZ, Mai YW, Liu S. Nanotechnology 2007;18:445602 (1–10).
214. Kashiwagi T, Harris Jr RH, Zhang X, Briber RM, Cipriano BH, Raghavan SR, et al. Polymer 2004;45:881–91.
215. Qin H, Zhang S, Zhao C, Hu G, Yang M. Polymer 2005;46:8386–95.
216. Kashiwagi T, Du F, Winey KI, Groth KM, Shields JR, Bellayer SP, et al. Polymer 2005;46:471–81.
217. Kashiwagi T, Grulke E, Hilding J, Harris R, Awad W, Douglas J. Macromol Rapid Commun 2002;23:761–5.
218. Ray SS, Bousmina M. Macromol Rapid Commun 2005;26:450–5.
219. Li Y, Shimizu H. Polymer 2004;45:7381–8.
220. Khatua BB, Lee DJ, Kim HY, Kim JK. Macromolecules 2004;37:2454–9.
221. Si M, Araki T, Ade H, Kilcoyne ALD, Fisher R, Sokolov JC, et al. Macromolecules 2006;39:4793–801.
222. Vo LT, Giannelis EP. Macromolecules 2007;40:8271–6.
223. Zhang Q, Yang H, Fu Q. Polymer 2004;45:1913–22.
224. Wang Y, Zhang Q, Fu Q. Macromol Rapid Commun 2003;24:231–5.
225. Hung HS, Hsu SH. Nanotechnology 2007;18:475101 (9 pp).
226. Chen CZ, Cooper SL. Adv Mater 2000;12:843–6.
227. Damm C, Mu¨ nstedt H, Ro¨sch A. Mater Chem Phys 2008;108:61–6.
228. Alt V, Bechert T, Steinru¨ cke P, Wagener M, Seidel P, Dingeldein E, et al. Biomaterials 2004;25:4383–91.
229. Hule RA, Pochan DJ. MRS Bull 2007;32:354–8.
230. Kim HW, Kim HE, Salih V. Biomaterials 2005;26:5221–30.
231. Huang J, Lin YW, Fu XW, Best SM, Brooks RA, Rushton N, et al. J Mater Sci Mater Med 2007;18:2151–7.
232. Lee YH, Lee JH, An IG, Kim C, Lee DS, Lee YK, et al. Biomaterials 2005;26: 3165–72.
233. Zhang Q, Zha L, Ma J, Liang B. Macromol Rapid Commun 2007;28:116–20.
234. Haraguchi K, Li HJ. Macromolecules 2006;39:1898–905.
235. Gupta AK, Gupta M. Biomaterials 2005;26:3995–4021.
236. Berry CC. J Mater Chem 2005;15:543–7.

237. Toprak MS, McKenna BJ, Waite JH, Stucky GD. Chem Mater 2007;19: 4263–9.
238. Vadala ML, Rutnakornpituk M, Zalich MA, Pierre St TG, Riffle JS. Polymer 2004;45:7449–61.
239. Rutnakornpituk M, Baranauskas VV, Riffle JS, Connolly J, Pierre St TG, Dailey JP. Eur Cells Mater 2002;3:102–5.
240. Rolland JP, Maynor BW, Euliss LE, Exner AE, Denison GM, Desimone JM. J Am Chem Soc 2005;127:10096–100.
241. Kongkanand A, Kuwabata S, Girishkumar G, Kamat P. Langmuir 2006;22: 2392–6.
242. DeLuca NW, Elabd YA. J Polym Sci Part B Polym Phys 2006;44:2201–25.
243. Kim YS,Wang F, Hickner M, Zadwodzinski TA, McGrath JE. J Membr Sci 2003; 212:263–82.
244. Zaidi SMJ, Mikhailenko SD, Robertson GP, Guiver MD, Kaliaguine S. J Membr Sci 2000;173:17–34.
245. Wang Z, Ni H, Zhao C, Li X, Fu T, Na H. J Polym Sci Part B Polym Phys 2006;44: 1967–78.
246. Li X, Xu D, Zhang G, Wang Z, Zhao C, Na H. J Appl Polym Sci 2007;103:4020–6.
247. Su YH, Liu YL, Sun YM, Lai JY, Guiver MD, Gao Y. J Power Sources 2006;155: 111–7.
248. Jiang R, Kunz HR, Fenton JM. J Membr Sci 2006;272:116–24.
249. Yang C, Srinivasan S, Arico AS, Creti P, Baglio V, Antonucci V. Electrochem Solid-State Lett 2001;4:A31–4.
250. Hill ML, Kim YS, Einsla BR, McGrath JE. J Membr Sci 2006;283:102–8.
251. Prashantha K, Park SG. J Appl Polym Sci 2005;98:1875–8.
252. Be´ bin P, Caravanier M, Galiano H. J Membr Sci 2006;278:35–42.
253. Thomassin JM, Pagnoulle C, Bizzari D, Caldarella G, Germain A, Je´ ro^me R. Solid State Ionics 2006;177:1137–44.
254. Barroso-Bujans F, Verdejo R, Arroyo M, Lopez-Gonzalez MM, Riande E, Lopez-Manchado MA. Macromol Rapid Commun 2008;29:234–8.
255. Baibarac M, Go´ mez-Romero P. J Nanosci Nanotechnol 2006;6:1–14.
256. Baughman RH, Zakhidov AA, De Heer WA. Science 2002;297:787–92.
257. Moniruzzaman M, Winey KI. Macromolecules 2006;39:5194–205.
258. Sandler JKW, Kirk JE, Kinloch IA, Shaffer MSP, Windle AH. Polymer 2003;44: 5893–9.
259. Bryning MB, Islam MF, Kikkawa JM, Yodh AG. Adv Mater 2005;17: 1186–91.
260. Kim YS, Wright JB, Grunlan JC. Polymer 2008;49:570–8.
261. Huynh WU, Dittmer JJ, Alivisatos AP. Science 2002;295:2425–7.
262. Liang Z, Dzienis KL, Xu J, Wang Q. Adv Funct Mater 2006;16:542–8.
263. Qi XY, Pu KY, Fang C,Wen GA, Zhang H, Boey FYC, et al. Macromol Chem Phys 2007;208:2007–17.

264. Olson DC, Shaheen SS, Collins RT, Ginley DS. J Phys Chem C 2007;111: 16670–8.

265. Kwong CY, Choy WCH, Djurisˇic´ AB, Chui PC, Cheng KW, Chan WK. Nanotechnology 2004;15:1156–61.

266. Kymakis E, Amaratunga GAJ. Appl Phys Lett 2002;80:112–5.

267. Tekin E, Smith PJ, Hoeppener S, van den Berg AMJ, Susha AS, Rogach AL, et al. Adv Funct Mater 2007;17:23–8.

268. Park JW, Ullah MH, Park SS, Ha CS. J Mater Sci Mater Electron 2007;18: S393–7.

269. Oey CC, Djurisˇic´ AB, Kwong CY, Cheung CH, Chan WK, Nunzi JM, et al. Thin Solid Films 2005;492:253–8.

270. Park JH, Lim YT, Park OO, Kim JK, Yu JW, Kim YC. Chem Mater 2004;16: 688–92.

271. Park JH, Lim YT, Park OO, Kim YC. Macromol Rapid Commun 2003;24: 331–4.

272. Park JH, Lim YT, Park OO, Kim JK, Yu JW, Kim YC. Adv Funct Mater 2004;14: 377–82.

273. Miyake J, Chujo Y. Macromol Rapid Commun 2008;29:86–92.

274. Stankovich S, Dikin DA, Dommett GHB, Kohlhaas KM, Zimney EJ, Stach EA, et al. Nature 2006;442:282–6.

275. Novoselov KS, Geim AK, Morozov SV, Jiang D, Zhang Y, Dubonos SV, et al. Science 2004;306:666–9.

276. Gilje S, Han S, Wang M, Wang KL, Kaner RB. Nano Lett 2007;7:3394–8.

277. Van Noorden R. Nature 2006;442:228–9.

278. Hatchett DW, Josowicz M. Chem Rev 2008. on web 01/03/2008.

279. Kong J, Franklin NR, Zhou C, Chapline MG, Peng S, Cho K, et al. Science 2000; 287:622–5.

280. An KH, Jeong SY, Hwang HR, Lee YH. Adv Mater 2004;16:1005–9.

281. Ali SR, Ma Y, Parajuli RR, Balogun Y, Lai WYC, He H. Anal Chem 2007;79: 2583–7.

282. Feng X, Mao C, Yang G, Hou W, Zhu JJ. Langmuir 2006;22:4384–9.

283. Zhang M, Fang S, Zakhidov AA, Lee SB, Aliev AE, Williams CD, et al. Science 2005;309:1215–9.

284. Vaia RA, Maguire JF. Chem Mater 2007;18:2736–51.

Chapter 8

ORGANO-SILOXANE SUPRAMOLECULAR POLYMERS USED IN CO_2 DETECTION-POLYMER THIN FILM

Gabriela Telipan[1], Lucian Pislaru-Danescu[1], Mircea Ignat[1], Carmen Racles[2]

[1]Research and Development National Institute for Electrical Engineering ICPE-CA

[2]Petru Poni Institute for Macromolecular Chemistry

INTRODUCTION

Supramolecular polymers are macromolecules in which the monomer units are kept together by non-covalent interactions. In a broader sense, the term is also used for self-organized macromolecules of conventional polymers involving non-covalent interactions to determine their material properties (M. Zigon, G. Ambrozic, 2003). It is well known that the physical properties of linear polymers and organic molecules are strongly modified when they contain associating end groups. Supramolecular polymers combine the features of supramolecular species with polymeric properties Self-organizing materials, which include liquid-crystalline polymers, block copolymers, hydrogen-bonded complexes, biological polymers, have great potential for various functional materials. Manipulation of supramolecular nanostructure in self-organizing materials is of critical importance for achieving desired functions and properties in solid state and liquid crystalline state molecular materials (M. Lee, B.K. Cho, et.al. 1999).

Although hydrogen bonds between neutral organic molecules are not very strong, they play a very important part in molecular recognition, self-assembling in bio-acromolecules, increasing miscibility in polymers blends, organization in liquid crystals. The directionality and versatility of the hydrogen bonds are the major keys for their implication in supramolecular polymers design. It has been observed that hydrogen bonding in supramolecular polymers is enhanced by liquid crystallinity and phase separation (. L. Brunsveld, B.J.B. Folmer, et.al., 2001).

The self-assembly of carboxylic acids as proton donors with pyridyl fragments as proton acceptors is most frequently used in the formation of H-bonded structures [4]. Such supramolecular polymers have been obtained and investigated, especially for liquid crystalline properties [M.Lee, B.K. Cho, et.al. 1999, M. Parra, P. Hidalgo et.al., 2005, H. an, A.H. Molla et.al., 1995, Y.S. Kang, H. Kim et.al., 2001, M. Parra, P. Hidalgo et. al., 2005; P.K. Bhowmik, X.Wang et. al., 2003, 2, 4-8]. Siloxane-containing benzoic acids and bipyridine also form interesting supramolecular structures, with cubic mesophases (E. Nishikawa, E. T. Samulski, 2000). This paper presents the sensitivity of gases in specially CO_2 of the organosiloxane supramolecular polymers.

SYNTHESIS AND CHARACTERIZATION OF ORGANO-SILOXANE SUPRAMOLECULAR POLYMERS

Materials

4,4`-Bipyridyne, (Fluka), m.p. 109-112oC, was used as received.

1,3-Bis(3-carboxypropyl)tetramethyldisiloxane, [HOOC(CH_3)3(CH3)2Si]2O, CX, was synthesized by using the method described in ref. [10] (the hydrolysis of 1,3-bis(3- cyanopropyl)-tetramethyldisiloxane); m.p. = 50 oC. IR, cm-1: 3000 cm+1 (OH), 1710 (-CO-); 1075 (Si-O-Si). 1HNMR (CCl4); ppm: 0.0-0.25 (Si-CH3); 0.41-0.85 (t, -CH2-Si); 1.4-2.0 (m, - CH2-); 2.15-2.6 (-CH2-COOH); 10.8-11.00 (COOH); 84% yield.

1,3-bis (carboxytrimellityliido-N-propylene) tetramethyl disiloxane, m.p. 200°C.

Methods for Polymers Synthesis

A series of organo-siloxane supramolecular polymers was obtained, starting from 4,4′- bipyridine as an acceptor and different silicon-containing carboxylic acids as hydrogendonor molecules [10].

The standard procedure was applied, i.e. mixing of stoechiometric amounts of 4,4′- bipyridine and siloxane diacids, in a non-polar solvent, followed by distillation and vacuum drying. We also tested the contact method, which gave very good results, proving that such supramolecular polymers are very easily obtainable, due to the natural molecular recognition process.

In this study, two of these supramolecular polymers were tested for potential application as gas sensors. The structures of the studied compounds are presented in Fig. 1. The starting disiloxane diacids, 1,3-bis(carboxypropyl)tetramethyl-disiloxane and 1,3-bis(carboxytrimellitylimido-N-propylene)tetramethyldisiloxane were synthesized following the methods described in the literature (J.E. Mulvaney, C.S. Marvel 1961, A. Staubli, E. Ron, R. Langer, 1990); (M.p. 50 °C).

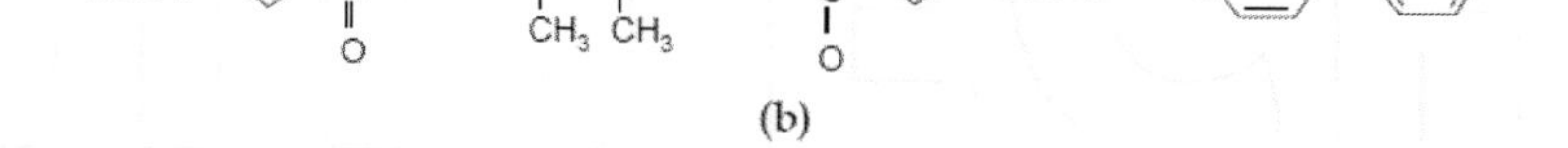

Figure 1. Chemical structure of the supramolecular polymer.(a): CH2 polymer; (b). CH5 polymer

Methods for Polymeric Structural Characterization of Polymer

Thermal Analysis

The thermal characterization by DSC showed for CH_2 a melting temperature of 63°C, while by polarized optical microscopy a narrow smectic mesophase was observed, between 62 and 68◦C. The heating rate didn't allow us to observe two endothermal peaks in DSC. For CH5, the DSC results confirmed the polarized optical microscopy observations, i.e. the presence of a mesophase between 130 and 160 ₒC, which were proved by the presence of two endothermal peaks on heating. In Fig. 2, the DSC curves of polymer CH5 are presented, at first (a) and second heating scan (b).

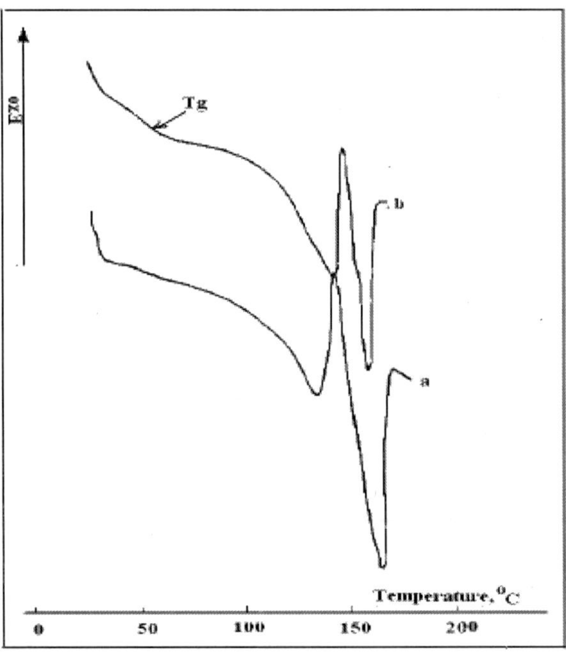

Figure 2. The DSC curves on first heating (a) diacid and the second heating (b) for the supramolecular supramolecular polymer CH5

FT-IR Spectra

The formation of the hydrogen bonds was primarily verified by FT-

IR spectroscopy which showed two specific broad absorption bands centered at around 1923 and 2475 cm-1 for CH_2 and at around 1940 and 2450 cm-1 for CH5 (H. Han, A.H. Molla, 1995), as can be observed in Fig. 3 and 4. All the other absorption bands, corresponding to the proposed structures were present: siloxane Si-O-Si asym. at 1030 cm^{-1}, CH3 def. at 1250 cm-1, CH_3 rock at 812 cm^{-1}, COOH at about 1700 cm-1, aromatic at 1600 cm-1. Nevertheless, a slight shift of the carboxylic C=O band did occur, reported to the starting diacids (see for example Fig. 3. All the other absorption bands, corresponding to the proposed structures were present in the polymers, with small shifts compared to the starting compounds, as can be observed in Table 1.

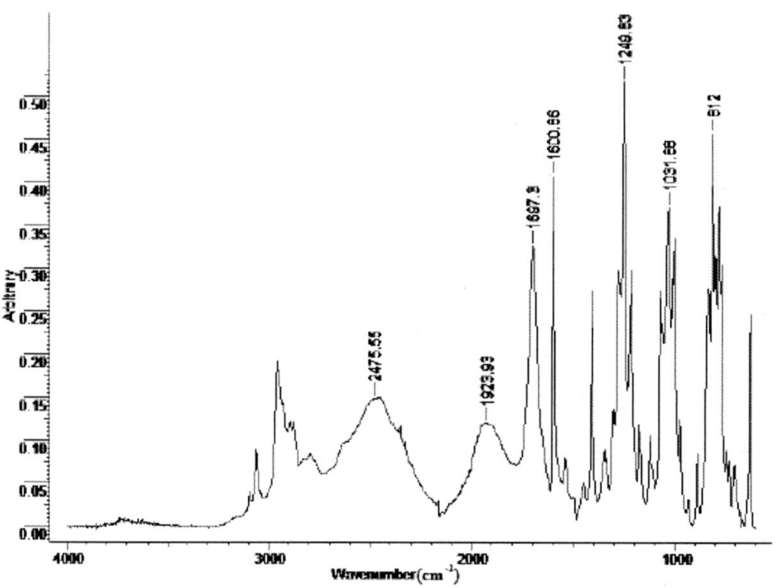

Figure 3. FT-IR spectra for CH_2 polymer of the supramolecular polymer

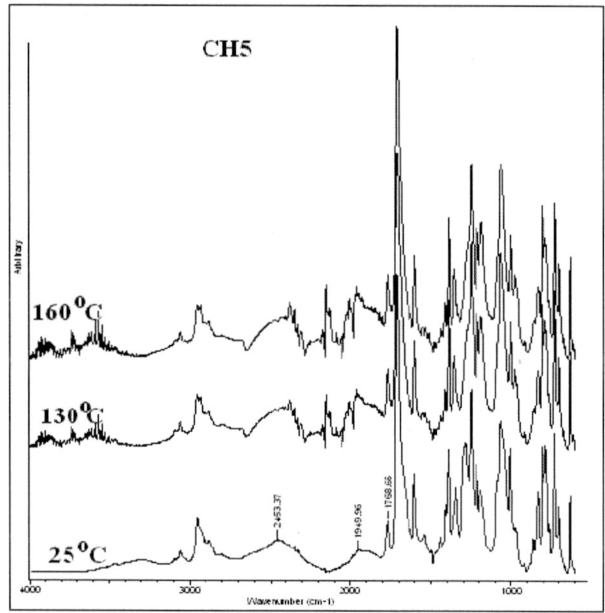

Figure 4. FT-IR spectra for CH_5 polymer

Table 1. The main FT-IR absorption bands in the starting compounds and supramolecular polymers

Compound	wavenumber, cm^{-1}	Assignment
4,4`-Bipyridine	1590	heteroaromatic
	812, 806	arC-H γ oop
1,3-bis(carboxytrimellityliido-N-propylene)tetramethyldisiloxane	787	arc-H 1,2,4 trisubstituted
	841, 1060, 1253	CH_3-Si; Si-O-Si
	1716	C=O acid
	1777	C=O imide
CH5	785	arc-H 1,2,4 trisubstituted
	807	arC-H γ oop
	838, 1067, 1252	CH_3-Si; Si-O-Si
	1600	aromatic
	1717	C=O acid
	1771	C=O imide
	1950, 2443	H-bonding

^{1}H-NMR Spectra

^{1}H-NMR spectra showed one set of signals and slightly modified chemical shifts compared to the starting compounds, indicating that the polymers were stable in chloroform, at least on the analysis time scale. The polymer spectrum is presented in Fig. 5.

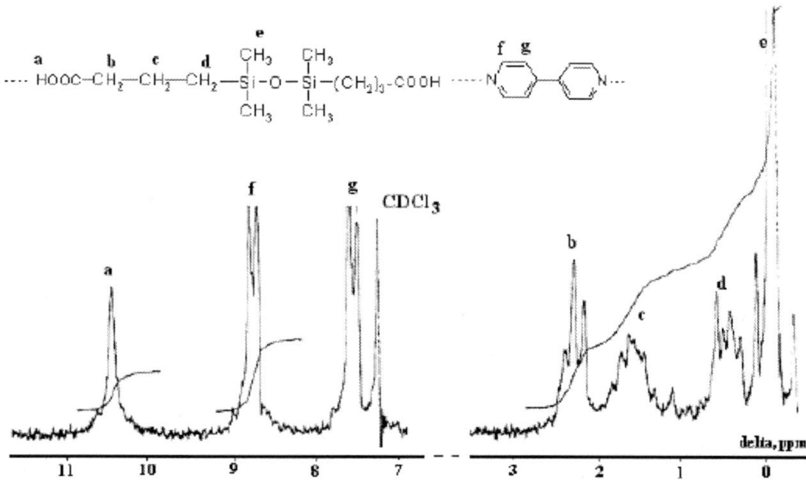

Figure 5. ^{1}H-NMR Spectrum of the supramolecular polymer

GAS SENSING MECHANISM OF ORGANO-SILOXANE SUPRAMOLECULAR POLYMERS

In order to analyse the influence of the testing gas - in this case CO_2 - on the chemical structure of the supramolecular polymers, we registered the IR spectrum of sample CH_2, after maintaining it for an hour in a CO_2 atmosphere. As can be observed in Fig. 6, the bands corresponding to the H-bond didn't change and no other existing absorption bands suffered noticeable modification. This result shows that the compound is chemically stable at exposure to the testing gas.

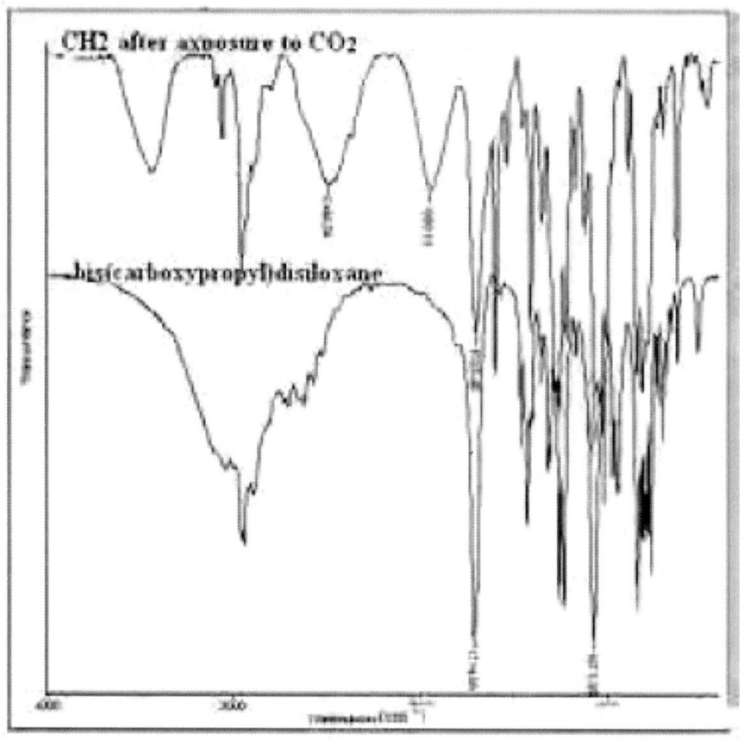

Figure 6. FT-IR spectra of the starting supramolecular polymer CH_5 before and after exposure to CO_2

CO_2 SENSING PROPERTIES OF POLYMERS

Thin and Thick Technology for Microsensors Fabrication

Device 1

Fig. 7 shows the structure of the sensor. An alumina subtrate 6 x 6 x 0.5 mm was used. On one side of the substrate was screen-printed an interdigitated electrode array using Au ink and heat treated at 950°C for 1 h. The sensitive layer obtained by dissolving the polymer in chloroform was deposed on the substrate with provided electrode by spin coating method. The sensor contains the pads of Pd - Ag conductive ink and the conductive layer of Ag ink heat treated at 750°C for 30 minutes.

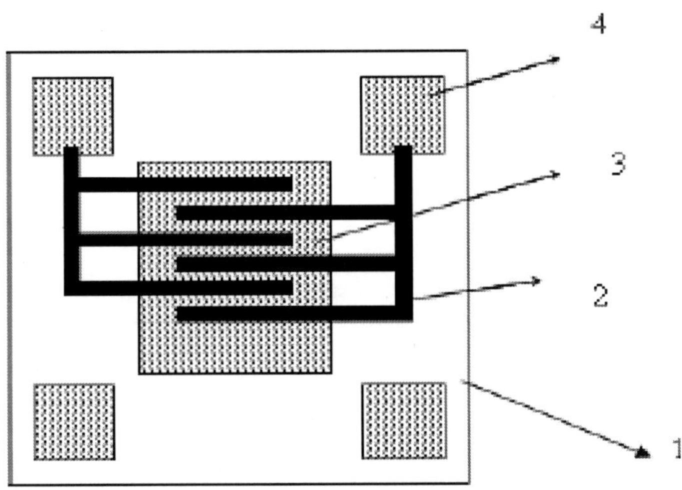

Figure 7. Sensor configuration; 1 - alumina substrate; 2 - electrode; 3 -polymeric layer; 4. pad.

Device 2

The sensor is composed from an alumina substrate 5 x 5 x 0.6 mm, where on one side of the substrate was deposed by magnetron sputtering 2 plates gold electrodes distanced at 3 mm. The deposed conditions were: pressure argon 1,9 torr, voltage 500V and the deposed time was 30 minutes. The sensitive layer obtained by dissolving the polymer in chloroform was deposed on the substrate with provided electrodes by spin coating method in the 200 nm thickness. The sensor was mounted on the transistor ambase-Fig. 8.

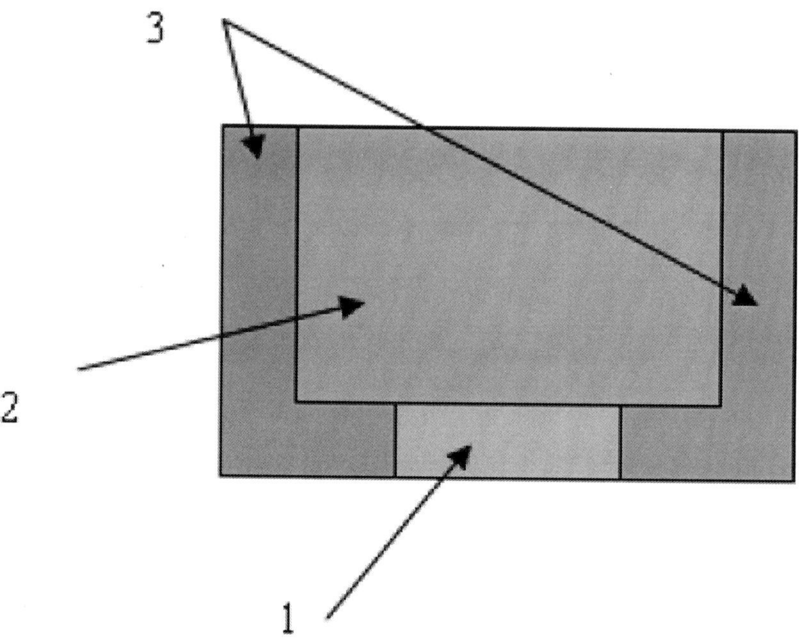

Figure 8. The sensor configuration 1- alumina substrate; 2- sensitive layer; 3- electrodes

The Characterization of Microsensors

Device 1

The Testing Measurement

The detection testing of gases CO_2 and NOx was performed in a 100 ppm concentration and the voltage measurements were effected with a multimeter type APPA 301 TRUE RMS. The sensor was exposed in the CO_2 and NOx atmospheres in 100 ppm concentration and was measured the voltage in function of the time. For the CH_2 polymer were obtained for CO_2 the voltage value about 70 mV after 10 minutes after gas exposure and for NOx was obtained the small value of the voltage by 4 mV. For the CH_5 polymer were obtained 238 mV for NOx and 92 mV for CO_2 atmospheres after 30 minutes exposure. Results that, the polymer CH_5 is more sensitive to gases against CH_2 polymer and the CH_5 polymer is more sensitive to NOx. CH_2 is sensitive to the CO_2 and unsensitive to NOx. Fig. 9 and 10 present the characteristics voltage - time for the CH_2 and CH_5 polymers.

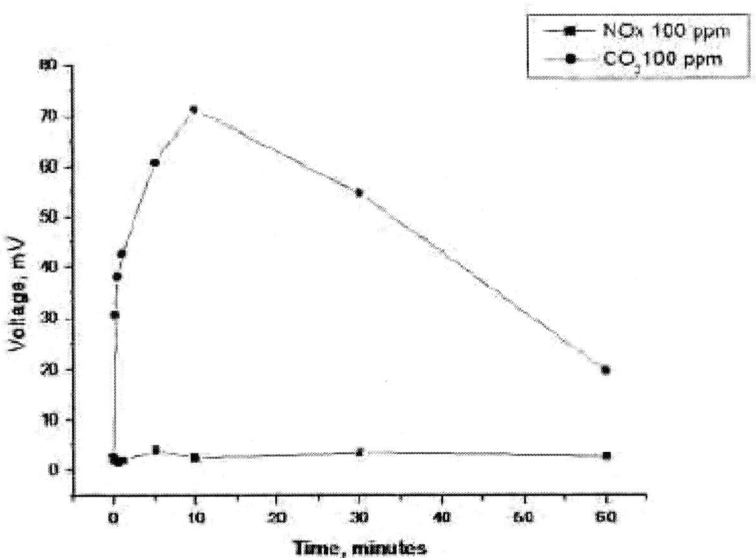

Fig. 9. The characteristic voltage time for the CH_2 polymer

Device 2

The Testing Measurements

The detection testing of gas CO_2 was performed in a 100 and 1000 ppm gas concentration and the voltage measurements were effected by testing module, in automated process mode. An control panel, provides a lot of measuring value, by rate 1/10 second.

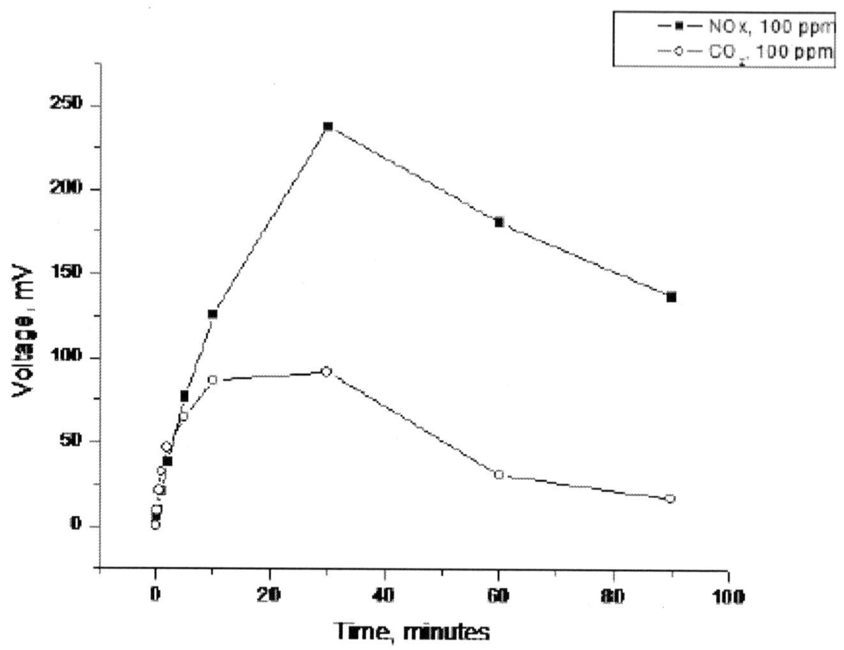

Figure 10. The characteristic voltage-time for CH_5 polymer

The bench of testing for the gas sensor consists in an enclosure where there are set the testing conditions of the sensor as well as in connected equipment. The whole process of testing is automated, being controlled by a programmable automaton. The gas for testing is introduced in a controlled way in the testing enclosure, through a mass debitmeter. In the testing enclosure is set a constant temperature, controlled by a temperature regulator-Fig. 11.

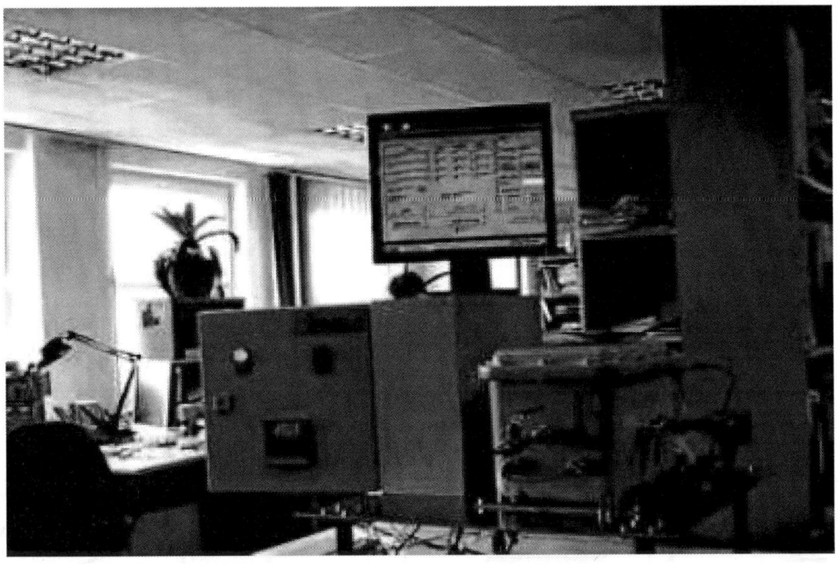

Figure 11. The apparatus for the gas sensor testing

The gas sensing characteristics The testing gas was CO_2 in 100 and 1000 ppm concentrations. Fig. 12 shows the characteristics voltage-time for 3 sensors exposed at 1000 ppm CO_2. It can be see that in the first 5 minutes gas exposure, the signals of the sensors are very weak with the voltage values of 660, 720 and 800 mV corresponding of sensor 1, sensor 2 and sensor 3. Fig. 13 presents the characteristics voltage-time for the sensor 3 exposed in 100 and 1000 ppm where were obtained the maximum voltage values of 92 mV for 100 ppm CO_2 and 970 mV for 1000 ppm CO_2 after 30 minutes gas exposure.

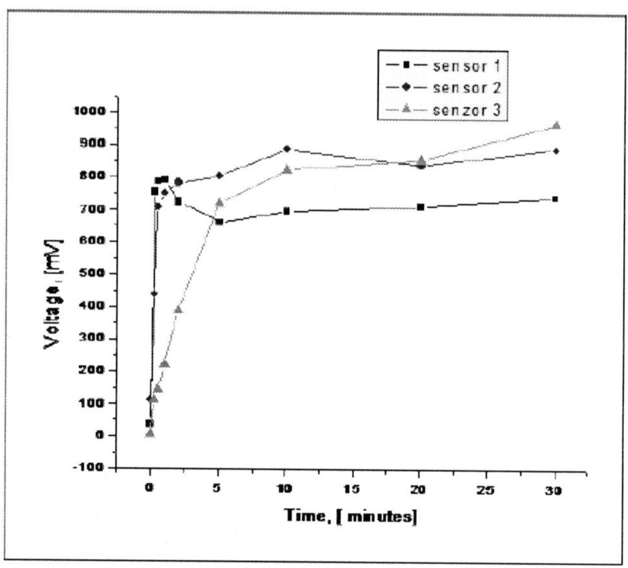

Figure 12. The characteristics voltage-time for 3 sensors exposed at 1000 ppm CO_2

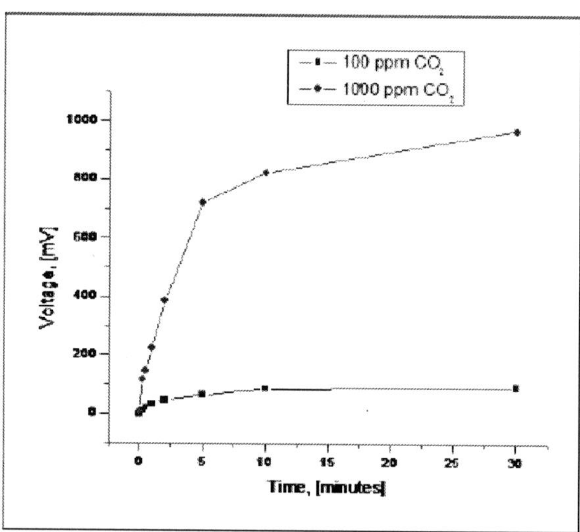

Figure 13. The characteristics voltage-time for sensor 3 exposed at 100 and 1000 ppm CO_2

The Signal Conditioning of Polymeric Sensitive Layer for Device 2

The electronic design of conditioning detector presented in Fig. 14, contains two amplification stages, cascade connected and realized with integrated circuit operational amplifier with JFET transistors at input, LF 356 type. The input impedance of these is theoretically infinite, practically very large, around Z ≈ 10MΩ. Therefore, the polarizing currents, when the sensitive element of the transducer is connected, are extremely small and the measured value is very precise. The sensitive element of the sensor of CO_2 is connected to the operational amplifier U1 in common mode, on the reversing input. Every one of these amplification stages realizes an amplification factor A_1 and A_2, respectively, corresponding with the realized reversing input.

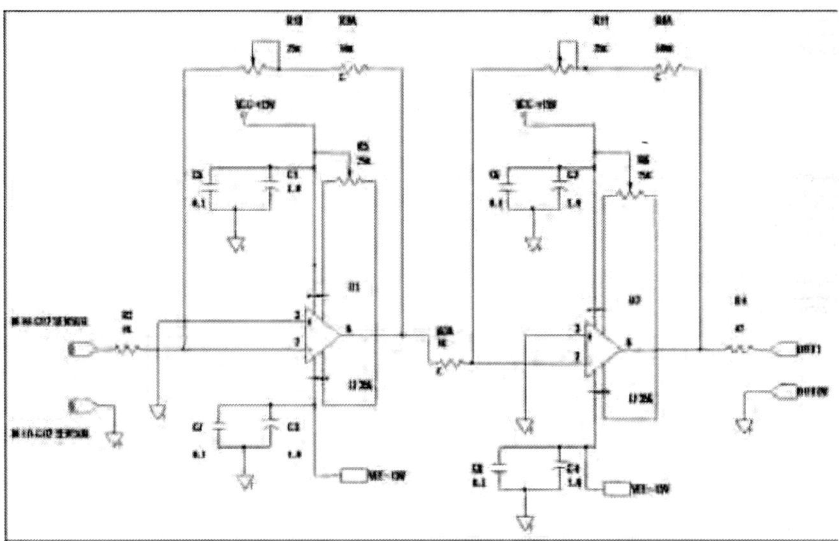

Figure 14. Electronic design of conditioning

The global amplification factor can be written, if we count that every amplification stage represents a basic cvadripole and the interaction is made exclusively through the terminals,

$$A = A_1 \times A_2 ,$$

where:

$$A1_{max} = -\frac{R9A + R10}{R3} = -\frac{10 + 25}{1} = -35, \quad (2)$$

and

$$A2_{max} = -\frac{R8A + R11}{R7A} = -\frac{100 + 25}{1} = -125. \quad (3)$$

We have obtained:

$$A_{max} = (-35)(-125) = 4375 \quad (4)$$

If the values of the potentiometers R_{10} and R_{11} are null, is obtained:

$$A_{min} = (-10)(-100) = 1000 \quad (5)$$

Obviously, if we would have wanted to use a single amplification stage for realizing the same maximal amplification value, most probably this would have transformed in an oscillator.

Of course, the final value of the global amplification is stabilized around the value A ≈ 2500, being related with the resulted values in the transfer characteristic of the sensitive element. The integrated circuit operational amplifiers U1 and U2, respectively, type LF 356, have capabilities of regulating the input offsets, through the R5 and R6 potentiometers. The regulating algorithm is:

With the reversing input of the operational amplifier U1 at the null potential, it is activating on the R5 potentiometer, until the output voltage, (pin 6 U1), is null, with R7A disconnected; With the reversing input of the operational amplifier U2 at the null potential, it is activating on the R6 potentiometer, until the output voltage, (pin 6 U2), is null, with R7A disconnected; In an iterative way, there are repeated the procedures in the stages 1 and 2, until the output voltage of the operational amplifier U2 is zero, when the reversing inputs of the operational amplifiers U1 and U2 are at null potential.

The output signal OUT1-Fig. 15, is a common mode voltage, picked at the output of the operational amplifier U2.

The supply voltage Vcc and VEE, respectively, is dual polarity power supply, stabilized. This is differentiated, ±15Vcc, in comparison with the null potential bar. The capacitor groups C1 – C5; C3 – C7;

C2 – C6; C4 – C8; make a decoupling of the supply voltage, in the immediate closeness of the operational amplifiers.

Another important design consideration is how circuit gain affects error, such as noise. When the operational amplifier U1 and U2 is operating at higher gains, the gain of the input stage is increased. At high gain, the input stage errors dominate. That is: input noise = eni and output noise = eno. Total noise RTI and RTO are calculated, (Charles Kitchin Lew Counts, 2004).

$$RTI = \sqrt{(eni)^2 + (eno / Gain)^2} \quad (6)$$

$$RTO = \sqrt{(Gain(eni))^2 + (eno)^2} \quad (7)$$

Typical noise of the LF 356 is specified, (LF356/LF357 JFET, 2001), as eni = 12nV / Hz , and eno = 80nV / Hz . Total noise RTI and RTO of the LF 356 operating at a gain of 10 is equal to:

$$RTI = 14{,}42 nV / \sqrt{Hz} \, , \; RTO = 144 nV / \sqrt{Hz}$$

Also, the noise input current of the LF 356 is very low, $0.01 pA / \sqrt{Hz}$.

Another design for this conditioning CO_2 transducer amplifier is showing in Fig. 15. The AD 8224 can operate on a ± 15 V dual supply. The most important parameter for this application is common - mode rejection ratio (CMRR). The AD 8224 has a high CMRR, of 94 dB for G = 10. In addition, slew rate of the AD 8224 is 2V / µs and the transfer function, (Charles Kitchin and Lew Counts, 2004).

$$G1 = 1 + \frac{49.4k\Omega}{R1A^*} \text{, respectively,} \quad G2 = 1 + \frac{49.4k\Omega}{R1B^*} \quad (8)$$

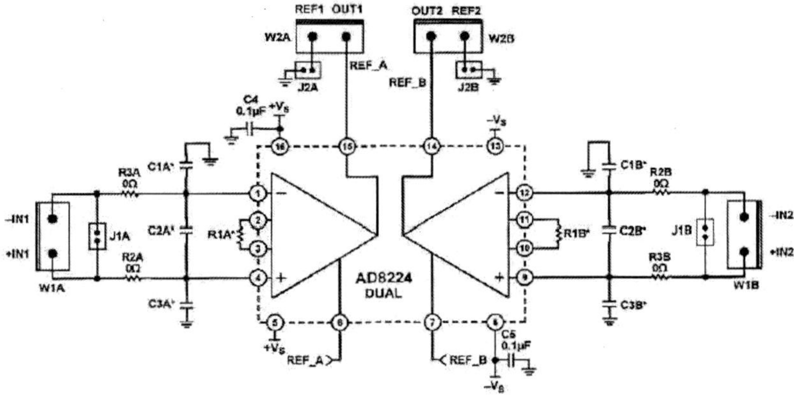

Figure 15. Precizion, AD 8224 JFET Input instrumentation amplifier for CO_2 transducer interface.

Placing a resistor across the R1A* and R1B* terminals, sets the gain of the AD 8224. For best performance, by using standard value of 1% resistor R1A* = R1B* = 5.49 KΩ, calculated gain is 9.998.

For optimum performance, a low - pass R - C network can be placed at the input of the instrumentation amplifier. This filter limits the input signal bandwidth according to the following relationship, (Charles Kitchin and Lew Counts, 2004):

$$FilterFreq_{DIFF} = \frac{1}{2\pi R2A\left(2C2A^{*} + C1A^{*}\right)} \quad (9)$$

where:

$$C1A^{*} = C3A^{*} \text{ and } R2A = R3A \quad (10)$$

As shown, Fig.15, R2A=R3A=4.02kΩ can be metal film resistors and C1A* = C3A* =1nF, C2A*= 10nF miniature size micas.

Fig. 16 presents the ansamble of image for CO_2 detector. As shown in Fig. 16, this application required a variable gain amplifier.

Figure 16. The image of CO_2 detector

THEORETICAL ASPECTS OF EQUIVALENT SCHEME

General Equivalent Circuit of Active Sensor

As a first step in the direction to make clear the study of gas sensor is following classification;

a. Active sensor and represents a sensor which convert the gas effect of the sensible material in a microvoltage -Fig. 17. Example: gas electrochemical sensor.

2. Passive sensor represents a sensor which convert the gas effect of the sensible material in a change of an electric resistance or an electric capacitance-Fig. 18,.example; gas semiconductor sensor.

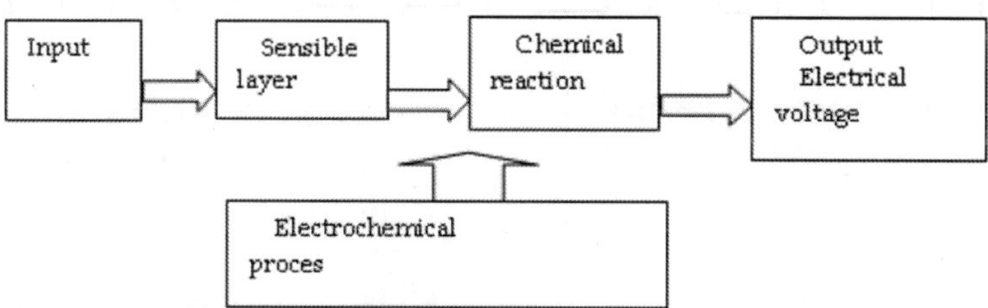

Figure 17. Active sensor

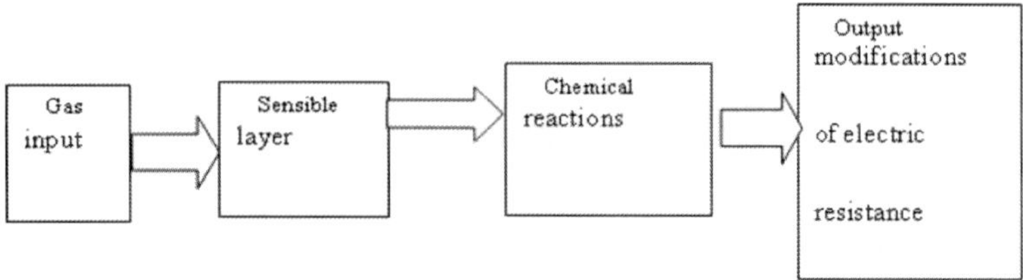

Figure 18. Pasive sensor

By the electric characteristic Gopel (W.Gopel, 1996), proposes a classification:

- Sensor type resistor.
- Sensor type diode.
- Sensor type capacitor.

An active gas sensor (with polymeric sensitive layer) accepting as a converter with stimuli gas concentration (percentage, ppm) and a terminal or electric port:voltage (U) is presented in fig.19.

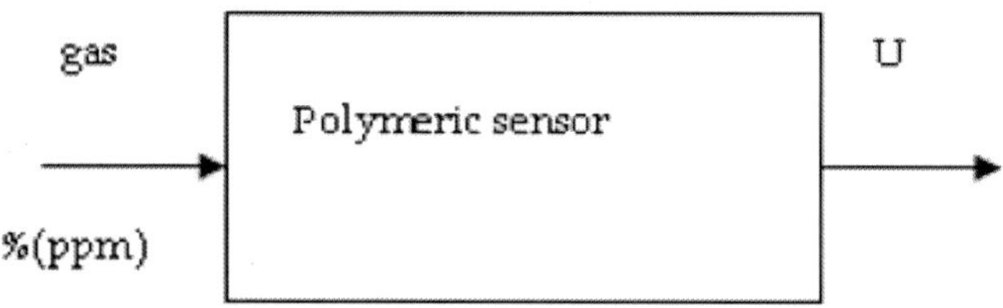

Figure 19. The gas sensor as a converter.

The Ideal Characteristics of an Equivalent Voltage Source

In keeping with the theory of electric circuits the active sensors have the following equivalent circuit -Fig. 20.

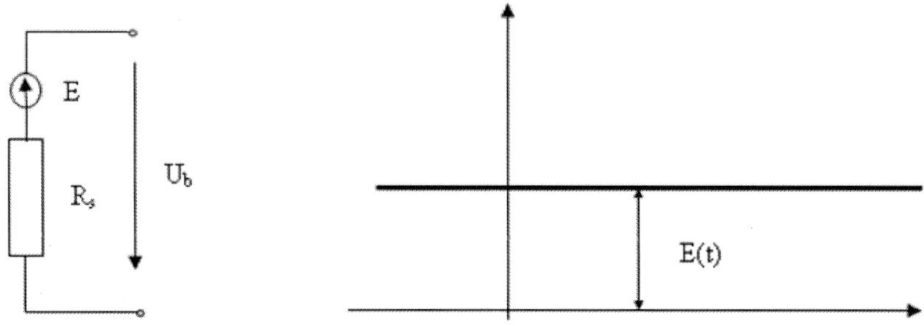

Figure 20. General equivalent circuit of active sensor

This equivalent circuit is a general scheme which represents a voltage source with base equations (Joubert) (C.I.Mocanu, 1979, C.A. Desoer,E.S.Kuh, 1969):

$$U_b - E = R_s i \qquad (11)$$

where E - the electromotive force, R_s voltage source series internal resistance, U_b terminal voltage of source, i - electric current.In Fig. 21 is presented and ideal characteristics for equivalent circuit.

Current Source Equivalent Source and Characteristics

Any voltage source consisting of an electromotive force E independent of the terminal current I , having an internal series resistances s R can be replaced by a current source

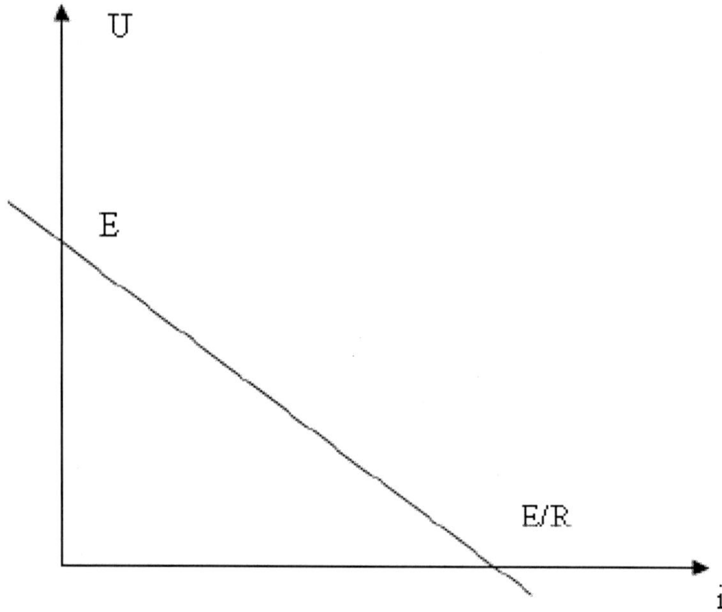

Figure 21. The ideal characteristics of an equivalent voltage source.

consisting of a current I $_g$ independent of the terminal voltage U $_b$, having an internal resistance g R with equation:

$$I_g - I = \frac{U_b}{R_g} \quad (12)$$

The behaviour of sensor gas asimilated as the voltage source is not linear on different subdomains and the real characteristics U(i) -Fig.22.

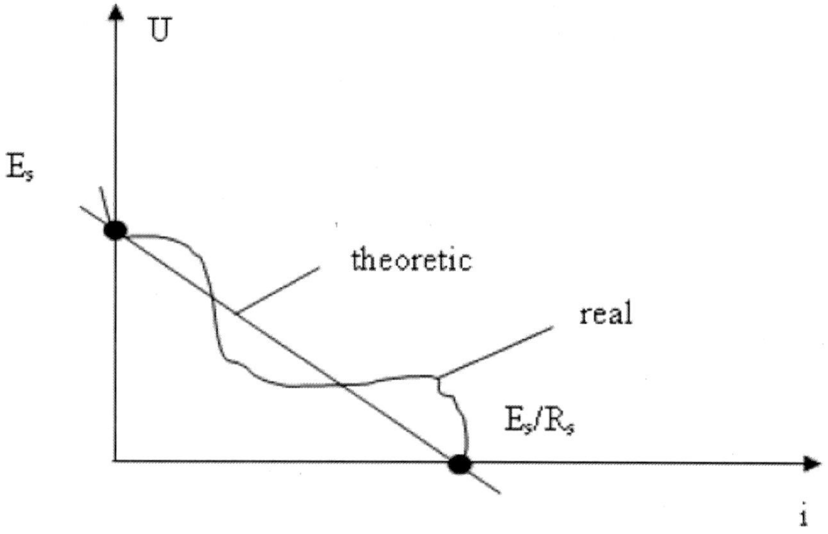

Figure 22. The real and ideal characteristics U(i) for a voltage source

Analog Equivalent Circuit of The Active Gas Sensor

The theoretical aspect which was presented forward is a synthetic macroscopic treatment of active sensor gas electric analysis. We consider for a detail study a geometric model –Fig. 23. Gaseous substances can affect the performance of the device. The adsorbtion and desorbtion such as gas molecules on the surface of sensible sensor layer may cause signal amplitude or phase or frequency fluctuation. This kind of perturbation can be described by dynamic process. The surface inference, fluctuations, particles adsorbtion, desorbtion and random movement of particles are the important aspects for the knowlidge of activ gas sensors.

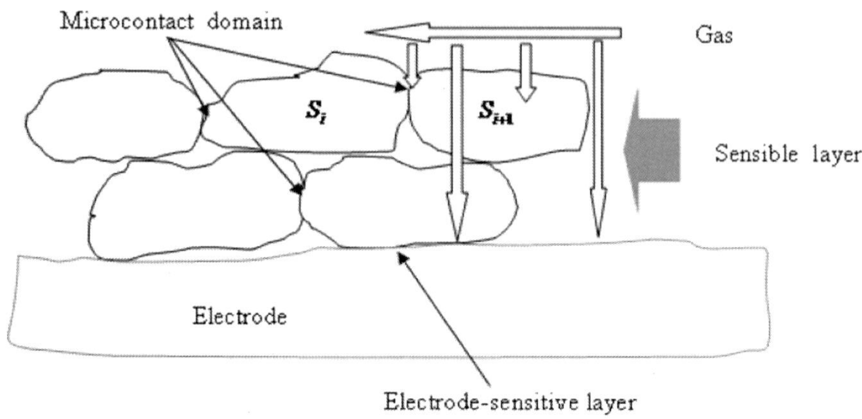

Figure 23. Sensor geometry modelling

Is necessary to formulate the electrokinetics problem of the active gas sensor:

Either a known topology of active gas sensor domain. To be investigated local and general equivalent circuit of domain referring to la imprintelectric field and the density (distributions) of this fields with the goal of modelling and determination of output electric signal; voltage, current and the specific parameters; sensibility, response time, linearity of signal, resolution, hysterezis.

Δ -active gas sensor domain is n-connected space; $\Delta = \Delta_s \cup \Delta_e \cup \Delta_{ec} \cup \Delta_{slc}$ and is formed by conductors (I and II category) in contact and which generate different imprint electric fields (C.A. Desoer,E.S.Kuh, 1969, C.I.Mocanu 1984):

- electrolityc nature with gas reaction $\overline{E}_{egr}$,
- electrolytic nature $\overline{E}_{ei}$ (another than gas sensor),
- microcontact nature , $\overline{E}_{ci}$

or others (C.A. Desoer, E.S. Kuh, 1969, C.I.Mocanu 1984). The domains represent; *sl* *sl* - domain of sensible layer, Δ *slc*-domain of microcontact between the different parts (subdomains) of sensible layers, Δe -domain of electrode, Δec -domain of microcontact between sensible layer and electrodes of sensor.

Either ϕ_g the molecular flux of the gas specie giving rise to the active sensor signal and referred to the geometrical model:

$$\Phi_g = \Phi_{react} + \Phi_{rm} = \Phi_{sl} + \Phi_e + \Phi_{ec} + \Phi_{slc} + \Phi_{rm} \quad (13)$$

where; ϕ_{sl} - the molecular flux to sensible part (layer) of active sensor, ϕ_e –the molecular flux to electrode surface, ϕ_{ec}- the molecular flux to the microcontact between electrode –sensible layer, ϕ_{slc}- the molecular flux to the contact between two sensible subdomains (S_i, S_{i+1}), Φ_{react} – the reacted flux molecules, Φ_{rm} – – the reflected molecular flux.

The sensible layer of sensor include $S_1, \dots, S_i, \dots S_n$ subdomains $\subset \Delta_{sl}$:

$$\Delta_{sl} = S_1 \cup S_2 \dots \cup S_i \cup S_{i+1} \cup \dots . \cup S_n. \quad (14)$$

The assumption of sensible layer subdomains $(S_1, \dots, S_i, \dots S_n)$ in microcontact is founded on the electron microscopie investigations (SEM) of sensible layer what indicates the existence of subdomains and the discontinuities in structure. We investigate especially the structure of NASICON ($Na_3Zr_2SiO_2PO_{12}$) disc, Na-super-ionic conductor/nitrite $TiO_2ZrO_2Y_2O_3$ and polymeric sensitive layers). We assume that each reacting molecules flux initiates a microvoltage source:

$$\Phi_{sl} - E_{sl},, \Phi_e - E_e, \Phi_{ec} - E_{ec}, \Phi_{slc} - E_{slc}$$

and the sensor domain can be modelled that in fig.24.

We assume that each reacting molecules flux initiates a microvoltage source:

$$\Phi_{sl} - E_{sl},, \Phi_e - E_e, \Phi_{ec} - E_{ec}, \Phi_{slc} - E_{slc}$$

and the sensor domain can be modelled that in fig.25.

In this circuit E, R represent the elements of voltage source; microelectromotive force (microvoltage) and electric resistance reported to the microsubdomains.

The detailed equivalent circuit includes a matrix structure (conform with the geometry and topology described - Fig.23) with elements of sensible subdomain in quadrature axis;

$$E_{sl_i}, R_{sl_i}, E_{sl_{i+1}}, R_{sl_{i+1}}$$ and in longitudinal axis;

$$E'_{sl_i}, R'_{sl_i}, E'_{sl_{i+1}}, R'_{sl_{i+1}}.$$

In diagram are evidenced the component possible noise signal of microcontact sensible layer- electrode(E_{ec}, R_{ec}) and reaction of gas with electrode (E_e, R_e) which have the longitudinal components and appear that effect of local microcorrosion.Essential signal for sensor is the microvoltage generated in sensible layer or sensible subdomain; sl E .

Because the detailed equivalent circuit requires many informations about the subdomain is proposes a synthetic circuit in Fig.25.

The electrokinetic equations by Kirkkof rules, are -Fig.24:

$$I_{sl} + I_{slc} = I_s \qquad (15)$$

For $(R_{sl}, E_{sl}) \,||\, (R_{slc}, E_{slc})$ the relations of microvoltage source and equivalent resistance are:

$$E_{eq} = \frac{E_{sl}R_{slc} + E_{slc}R_{sl}}{R_{sl} + R_{slc}} \tag{16}$$

$$R_{eq} = \frac{R_{sl}R_{slc}}{R_{sl} + R_{slc}} \tag{17}$$

and the equivalent circuit becomes that in Fig 26 with the general equation;

$$(E_{eq} + E_{ec} + E_{c}) - U_{s} = I_{s}(R_{eq} + R_{ec} + R_{e}) \tag{18}$$

Finnaly:

$$U_{s} = (E_{eq} + E_{ec} + E_{e}) - I_{s}(R_{eq} + R_{ec} + R_{e}) \tag{19}$$

or with equivalent relations of microvoltage source and resistance:

$$U_{s} = \frac{E_{sl}R_{slc}}{R_{sl} + R_{slc}} + \frac{E_{slc}R_{sl}}{R_{sl} + R_{slc}} + E_{ec} + E_{e} - I_{s} \cdot \frac{R_{sl}R_{slc} + (R_{ec} + R_{e})(R_{slc} + R_{sl})}{R_{sl} + R_{slc}}$$

There are following cases:

I. $E_{slc} = E_{ec} = E_{e} = 0$ represents the ideal case when the are nor internal noise microvoltage sources:

$$U_{s} = E_{sl}k_{1} - k_{2} \tag{20}$$

where:

$$k_{1} = \frac{R_{slc}}{R_{sl} + R_{slc}}, k_{2} = \frac{R_{sl}R_{slc} + (R_{ec} + R_{e})(R_{slc} + R_{sl})}{R_{sl} + R_{slc}} \cdot I_{s} \tag{21}$$

The characteristic $U_{s}(E_{s})$ is presented in Fig.27.

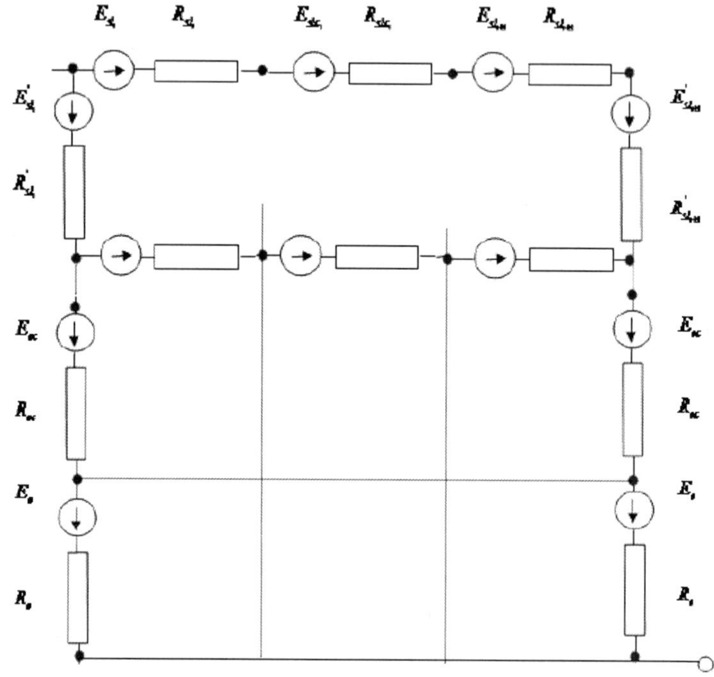

Figure 24. Detailed equivalent circuit of active gas sensor

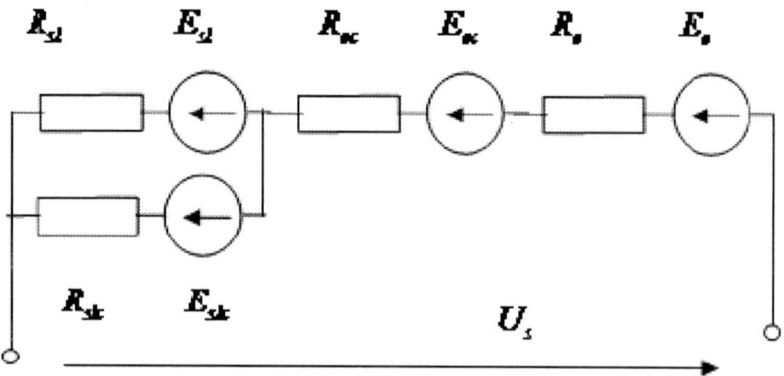

Figure 25. An equivalent synthetic circuit of active gas

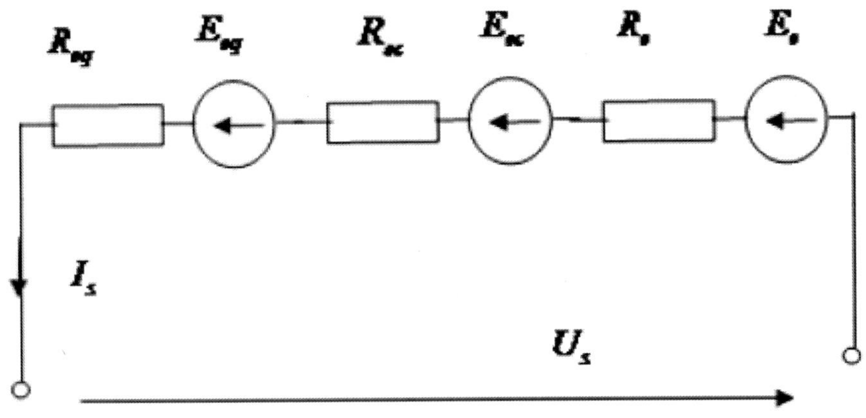

Figure 26. The final equivalent synthetic circuit of active gas

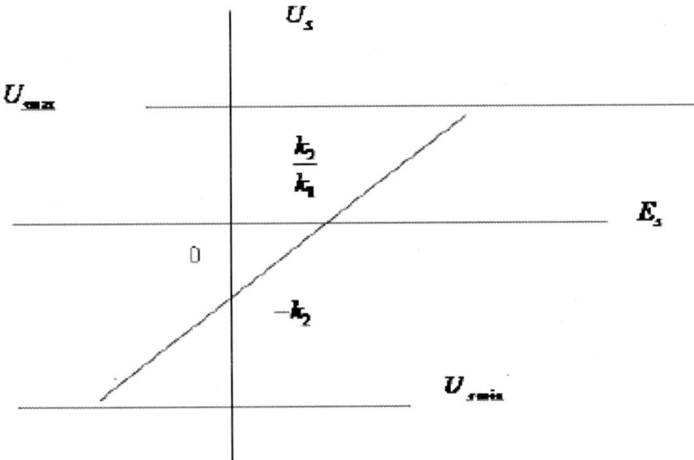

Figure 27. Characteristic U_s (E_s) in ideal case without the noise microsources

Essential every problem of gas sensor generation; measurement, control, involves theconversion of the conversion of the chemical in electrical energy. In dealing, with active gas structure, chemical-electric coupled systems, it is very convenient at some times and essential at others to treat the combined gas sensor as an entity. The basic idea of an electrochemical analogy comparison is the following:

Table 2. Basic analogies

	Gas circuit	Electric circuit
Potential	Concentration - c	Electric potential - V
	$\Delta c = c_2 - c_1$ Concentration difference	$\Delta U = V_2 - V_1$ Voltage (Potential difference)
Flux density	$\Phi_d = -DS\frac{\partial c}{\partial x}$, $\Phi_{dg} = \frac{\Phi_d}{S} = -\frac{\nabla c}{\frac{1}{D}}$ Fick's difussion flux dernsity S-surface of diffusion	$\bar{J} = \sigma\bar{E}, I = \frac{U}{R}$ Ohm's law $\bar{J}$ – electric density current, I – electric current, R – electric resistance
Constitutive parameter	D – diffusion coefficient	σ – electric conductivity

According to Fick's first diffusion law [5.5] the evolution of the gas in active sensor, the rate (flux) Φ_d depends on the concentation gradient c, on sensible material subdomain S_i:

$$\Phi_{S_i} = \Phi_d = -DS\left(\frac{dc}{dx}\right)_{x=\frac{d}{2}} \tag{22}$$

where S is the surface area and D the diffusion coefficient, d- the thickness of sensible subdomain. To obtain the gas evolution rate as a function of gas active sensor, the time dependence of dc/dx must be known. By Fick's second diffusion law:

$$\frac{\partial c}{\partial x} = D\frac{\partial^2 c}{\partial x^2} \tag{23}$$

If 0 dc = c_0 dp with 0 c-initial concentration and p - gas pression, according to Henry law equation may therefore be written as:

$$\Phi_p = DSc_o(\frac{dp}{dx}) \tag{24}$$

which represent another analog model with gas pression p -potential.

Is possible to determine the solute distributions c(x,t) during transient states, (I. Lundsrom, 1996, W. Jost, 1952, H. Carslaw, J. C. Jaeger, 1959, V. S. Vladimirov, 1980, J. Crank, 1956).

In the case of uniform initial concentration equation (25) can be satisfied by the solution (V.S. Vladimirov, 1980):

$$c(x,t) = C_i \frac{4}{\pi} \sum_{n=0}^{\infty} \frac{(-1)^n}{2n+1} \cos \frac{\pi x(2n+1)}{d} \cdot \exp\left\{ -\left[\pi(2n+1)\right]^2 \frac{Dt}{d^2} \right\} \tag{25}$$

where t --time and i C -- initial gas concentration.

With the basic analogies which was described in Table 1, is presented in Fig. 28 the equivalent circuit of the active gas sensor as a chemical gas subsystem (input of sensor) coupled with an electric subsystem (output of the sensor).

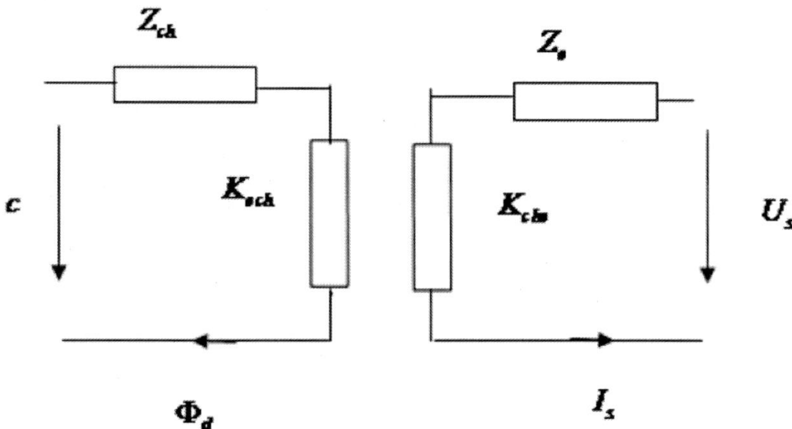

Figure 28. Analog equivalent circuit of the active gas sensor.

The equation of this coupled circuits are:

Input

$$c = K_{che} I_s + Z_{ch}^{I_s} \Phi_d \tag{26}$$

Output

$$U_s = Z_e^{\Phi_d} I_s + K_{ech} \Phi_d \tag{27}$$

where c -- gas concentration, $U_s -$ output voltage, $\Phi_d -$ flux gas, I_s --electric current sensor, $Z_{ch}^{I_s} -$ specific chemical gas sensor impedance to $I_s = 0$, (no load output):

$$Z_{ch}^{I_s} = \frac{c}{\Phi_{d\,(I_s=0)}} \tag{28}$$

$Z_e^{\Phi_d} -$ specific electric impedance to $\Phi_d = 0$ (no-load input):

$$Z_e^{\Phi_d} = \frac{U_s}{I_{s\,(\Phi_d=0)}} \tag{29}$$

$K_{che} -$ coupled ratio or operator between chemical gas input and electric output:

$$K_{che} = \frac{c}{I_{s\,(\Phi_d=0)}} \tag{30}$$

K_{ech} - coupled ratio or operator between the electric output and chemical gas input:

$$K_{ech} = \frac{U_s}{\Phi_{d\ (I_s=0)}} \quad (31)$$

The coupled operators; K_{che} and K_{ech} represents the reflected impedances.

CONCLUSIONS

A organo-siloxane supramolecular polymer was obtained, starting from 4,4′-bipyridine as an acceptor and silicon-containing carboxylic acids as hydrogen-donor molecules. The sensors were made by thin and thick film technologies. The thickness of polymeric layer deposed by spin coating on the alumina substrate was 200 nm. For the device 1, the sensor was exposed in the CO_2 and NOx atmospheres in 100 ppm concentration gas and was measured the voltage in function of the time. For the CH_2 polymer were obtained for CO_2 the voltage value about 70 mV after 10 minutes after gas exposure and for NOx was obtained the small value of the voltage by 4 mV. For the CH5 polymer were obtained 238 mV for NOx and 92 mV for CO_2 atmospheres after 30 minutes exposure. For the 3 sensors of the device 2, exposed at 1000 ppm CO_2, were obtained in the first 5 minutes gas exposure very weak voltage signals of 660, 720 and 800 mV corresponding of sensor 1, sensor 2 and sensor 3. For the sensor 3 exposed in 100 and 1000 ppm CO_2 were obtained the maximum voltage values of 92 mV for 100 ppm CO_2 and 970 mV for 1000 ppm CO_2 after 30 minutes gas exposure. The apparatus for CO_2 detection is composed by the sensor and electronic device of signal conditioning. In the paper were discussed the theoretical aspects about equivalent scheme circuit of active sensors.

REFERENCES

1. P.K. Bhowmik, X.Wang, H. Han, Main-chain, thermotropic, liquid-crystalline, hydrogenbonded polymers of 4,4'-bipyridyl with aliphatic dicarboxylic acids J. Polym. Sci., Part A: Polym. Chem., 2003, vol. 41 (9), pp. 1282-1295
2. L. Brunsveld, B.J.B. Folmer, E.W. Meijer, R.P. Sijbesma,Supramolecular polymers, Chem. Rev.2001, vol. 101, pp.4071.
3. H.Carslaw, J.C.Jaeger "Conduction of Heat in Solids",Oxford University Press,Fair Law, 1959.
4. C.A. Desoer, E.S. Kuh "Basic Circuit Theory" New York, Editions,Mc.Graw Hill,1969. W.Gopel "Ultimate limits in the miniaturization of chemical sensors", Sensors and Actuators, A56 (1996), pp. 83-102.
5. H. Han, A.H. Molla, P.K. Bhowmik, Hydrogen-Bonded Main-Chain Liquid Crystalline Polymers of Trans-1,2-Bis(4-pyridyl)ethylene with Aliphatic Dicarboxylic Acids. Polym. Prepr., Am. Chem. Soc. Div. Polym. Chem. 1995, 36(2), 332-333.
6. Y.S. Kang, H.Kim, W.C. Zin, Phase behaviour of hydrogen-bonded liquid crystalline complexes of alkoxycinnamic acids with 4,4'-bipyridine, Liquid Crystals, Volume 28, Number 5, 1 May 2001, pp. 709-715.
7. Charles Kitchin and Lew Counts, "A Designer's Guide to Instrumentation Amplifiers, Analog Devices, Inc. Printed in U.S.A, 2ND Edition, 2004, pp. 64-74.
8. W.Jost "Diffusion in Solids Liquids,Gases" Academic Press Inc., New York, 1952.
9. M.Lee, B.K. Cho, Y.S. Kang, W.C. Hydrogen-Bonding-Mediated Formation of Supramolecular Rod-Coil Copolymers Exhibiting Hexagonal Columnar and Bicontinuous Cubic Liquid Crystalline Assemblies. Zin, Macromolecules, 1999, vol. 2, pp. 8531-8537.
10. LF356/LF357 JFET Input Operational Amplifiers, National Semiconductor Corporation, 2001.
11. I.Lundsrom "Approaches and mechanisms to solid state based sensing", Sensor and Actuators B35-36 (1996), pp. 11-19.
12. C.I.Mocanu "Teoria circuitelor electrice"Ed.Didactica si Pedagogica, Bucuresti,1979. C.I.Mocanu"Teoria câmpului electromagnetic", Ed.Didactică şi Pedagogică, Bucureşti, 1984.
13. J.E. Mulvaney, C.S. Marvel, J. Polym. Sci., 1961, Vol. 50, pp. 41.
14. E. Nishikawa, E. T. Samulski, New mesogens with cubic phases: hydrogen-bonded bipyridines and siloxane-containing benzoic acids I. Preparation and phase behaviour, Liq. Cryst., 2000, vol. 27(11), pp. 1457-1462.
15. M. Parra, P. Hidalgo, J. Alderete, New supramolecular liquid crystals induced by hydrogen bonding between pyridyl-1,2,4-oxadiazole derivatives and 2,5-thiophene dicarboxylic acid, Liq. Cryst., 2005, Vol. 32(4), pp. 449-455.
16. M. Parra, P. Hidalgo, J. Barbera, J. Alderete, Properties of thermotropic liquid

crystals induced by hydrogen bonding between pyridyl-1,2,4-oxadiazole derivatives and benzoic acid, 4-chlorobenzoic acid or 4-methylbenzoic acid, Liq. Cryst., 2005, vol. 32(5), pp. 573-577.

17. A. Staubli, E. Ron, R. Langer, Hydrolytically Degradable Amino Acid Containing Polymers. J. Am. Chem. Soc., 1990, Vol. 112, pp. 4419.

18. V.S.Vladimirov "Ecuatiile fizicii matematice" Ed.Stiintifica si Enciclopedica, Bucuresti,1980.

19. M. Zigon, G. Ambrozic, Supramolecular polymers = Supramolekularni polimeri Materiali in Tehnologije, 2003, vol 37 (5), pp. 231-236.

Citations

CHAPTER 1

C.C. Wang, G. Pilania, S.A. Boggs, S. Kumar, C. Breneman, R. Ramprasad Computational Strategies for Polymer Dielectrics Design http://dx.doi.org/10.1016/j.polymer.2013.12.069

CHAPTER 2

Nanopolymers improve delivery of exon skipping oligonucleotides and concomitant dystrophin expression in skeletal muscle of mdx mice BMC Biotechnology 2008, 8:35 doi:10.1186/1472-6750-8-35.

CHAPTER 3

Jeffrey R. Potts, Daniel R. Dreyer, Christopher W. Bielawski, Rodney S. Ruoff, Graphene-based polymer nanocomposites, http://dx.doi.org/10.1016/j.polymer.2010.11.042

CHAPTER 4

Gérrard Eddy Jai Poinern1 Xuan Thi Le Songhua Shan Trevor Ellis Stan Fenwick John Edwards Derek Fawcett Ultrasonic synthetic technique to manufacture a pHEMA nanopolymeric-based vaccine against the H6N2 avian influenza virus: a preliminary investigation published online Sep 28, 2011. doi: 10.2147/IJN.S24272.

CHAPTER 5

V. Tamara Perchyonok, Shengmiao Zhang and Theunis Oberholzer Protective Effect of Conventional Antioxidant (!-Carotene, Resveratrol and Vitamin E) in Chitosan-Containing Hydrogels Against Oxidative Stress and Reversal of DNA Double Stranded Breaks Induced by Common Dental Composites: In-Vitro Model [DOI: 10.2174/1874140101307010001].

CHAPTER 6

M. Essalhi, M. Khayet, C. Cojocaru, M.C. García-Payo and P. Arribas Response Surface Modeling and Optimization of Electrospun Nanofiber Membranes The Open Nanoscience Journal 02/2013; 7(1):8-17. DOI:10.2174/1874140101307010008

CHAPTER 7

D.R. Paul, L.M. Robeson, Polymer Nanotechnology: Nanocomposites http://dx.doi.org/10.1016/j.polymer.2008.04.017

CHAPTER 8

Gabriela Telipan, Lucian Pislaru-Danescu, Mircea Ignat, Carmen Racles (2010). Organo-Siloxane Supramolecular Polymers Used in CO2 Detection-Polymer Thin Film, Polymer Thin Films, Abbass A Hashim (Ed.), ISBN: 978-953-307-059-9, InTech, DOI: 10.5772/8398.

INDEX